METHODS IN MOLECULAR BIOLOGY

Series Editor:
John M.Walker
University of Hertfordshire
School of Life and Medical Sciences
Hatfield, Hertfordshire, AL10 9AB, UK

For further volumes:
http://www.springer.com/series/7651

NLR Proteins

Methods and Protocols

Edited by

Francesco Di Virgilio

*Department of Morphology, Surgery and Experimental Medicine,
University of Ferrara, Ferrara, Italy*

Pablo Pelegrín

*Molecular Inflammation Group, Murcia Biomedical Research Institute (IMIB-Arrixaca),
Hospital Virgen de la Arrixaca, Murcia, Spain*

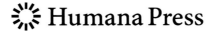

Editors
Francesco Di Virgilio
Department of Morphology
Surgery and Experimental Medicine
University of Ferrara
Ferrara, Italy

Pablo Pelegrín
Molecular Inflammation Group
Murcia Biomedical Research Institute (IMIB-
    Arrixaca), Hospital Virgen de la Arrixaca
Murcia, Spain

ISSN 1064-3745   ISSN 1940-6029   (electronic)
Methods in Molecular Biology
ISBN 978-1-4939-3564-2   ISBN 978-1-4939-3566-6   (eBook)
DOI 10.1007/978-1-4939-3566-6

Library of Congress Control Number: 2016935501

Printed on acid-free paper

This Humana Press imprint is published by Springer Nature
The registered company is Springer Science+Business Media LLC New York

Preface

Inflammation is the most fundamental defense mechanism developed by multicellular organisms [1]. Central to inflammation is the ability to discriminate noxious from non-noxious agents and to detect signs of tissue damage or cellular distress that might signal an impending danger. Given the multiplicity of foreign microorganisms which our body gets in contact with throughout its life, the ability to tell "dangerous" from "non-dangerous" is crucial because we do not want to rouse a potentially destructive response, such as inflammation, if not absolutely necessary. The immune system selected the capability to cause cell or tissue damage as an unequivocal proof of a given microorganism "dangerousness." Furthermore, as it is well known to clinicians, harmful agents are very often of endogenous origin (e.g., misfolded proteins or products of abnormal metabolic pathways); thus, an efficient defense mechanism should also be able to detect the presence of these agents. Thus, as pointed out by Carl Nathan, in order to start inflammation in response to a foreign microorganism, our body needs to detect the guest (the pathogen) and have evidence of its "dangerousness," i.e., detect the damage [2]. This "two-hit" system is based on the ability to identify on one hand molecular signs of the presence of the pathogen, i.e., pathogen-associated molecular patterns, PAMPs, and on the other molecular signs of possible cell damage or distress, i.e., damage-associated molecular patterns, DAMPs [3]. Very interestingly, DAMPs released as a consequence of sterile tissue damage, as, for example, in the case of closed trauma, autoimmune diseases, or metabolic stress, are themselves sufficient to trigger inflammation (sterile inflammation), in the absence of PAMPs. This self-sufficiency of endogenous factors may tell us something of the evolutionary driving forces behind the inflammatory response.

Immune cells have developed a sophisticated array of receptors to monitor the extracellular and intracellular environment for the presence of PAMPs and DAMPs. At least four different families of receptors are known: (1) C-type lectin receptors (CLRs), (2) retinoid-acid inducible gene (RIG)-I-like receptors (RLRs), (3) toll-like receptors (TLRs), and (4) nucleotide-binding oligomerization domain (NOD)-like receptors (NLRs). TLRs are specialized to detect pathogens in the extracellular space or in the endosome lumen that can be equated to the extracellular space. NLRs are specialized to sense pathogen presence in the cytoplasm; thus, they can be considered the prototypic intracellular pathogen/danger-sensing receptors. Over the last 10 years, our knowledge of NLRs, as well as the number of diseases in which these key defense molecules are involved, has increased exponentially. We now know that some NLRs are intracellularly assembled together with other proteins (i.e., ASC and caspase-1) in a macromolecular complex named the inflammasome and that this complex undergoes a complex molecular rearrangement during immune cell activation [4]. Such changes (translocation to different intracellular compartments, shift in the affinity for intracellular nucleotides, conformational changes) are intimately linked to immune cell effector responses.

The tumultuous growth of interest on the NLRs requires a parallel increase in the technical weaponry for the molecular and biochemical investigation. This is the need that this book aims to satisfy. In the first chapter (by Edward Lavelle), we provide a succinct albeit authoritative appraisal of current knowledge of innate immune receptors. In the second chapter Fayyaz Sutterwala reports on the so-called "atypical" inflammasomes. Then four chapters, by Isabelle Couillin, David Brough, Pablo Pelegrin, and Veit Hornung, follow on classical biochemical and novel bioluminescence techniques for the measurement of IL-1β release as a readout of inflammasome activation. Francesco Di Virgilio describes a novel bioluminescent probe for in vivo imaging of extracellular ATP, the prototypic DAMP. The following five chapters, by Christian Stehlik, Bernardo Franklin, Fatima Martin, Monica Comalada, and Ming-Zong Lai, reports on different biochemical and microscopy techniques that can be used to monitor NLR oligomerization. The following chapters focus on the consequences of inflammasome activation. Virginie Petrilli describes techniques to measure caspase-1 activation. Fabio Martinon reports on cell-free systems for the study of inflammasome function. On the other hand, Vincent Compan details the protocols for NLR reconstitution in a cell model such as HEK293 cells. Gloria Lopez-Castejon describes the procedure to investigate posttranscriptional NLR modifications. Lorenzo Galluzzi describes the protocols for the study of one of the most unfavorable consequences of inflammasome activation, i.e., cell death. *Finally, Marco Gattorno and Anna Rubartelli give an update appraisal of the application of inflammasome studies to the clinic.*

We hope that this book will provide a sound basis for the molecular investigation of NLR function in health and disease and will sparkle interest in these fascinating molecules by investigators from many different and faraway disciplines.

Ferrara, Italy   *Francesco Di Virgilio*
Murcia, Spain   *Pablo Pelegrín*

References

1. Medzhitov R (2008) Origin and physiological roles of inflammation. Nature 454: 428–435
2. Nathan C (2002) Points of control in inflammation. Nature 420:846–852
3. Schroder K, Tschopp J (2010) The inflammasomes. Cell 140:821–832
4. Gross O, Thomas CJ, Guarda G, Tschopp J (2011) The inflammasome: an integrated view. Immunol Rev 243:136–151

Contents

Contributors

JAMES BAGNALL • *Faculty of Life Sciences, University of Manchester, Manchester, UK*

EVA BARTOK • *Institute of Molecular Medicine, University Hospital, University of Bonn, Bonn, Germany; Institute of Clinical Chemistry and Clinical Pharmacology, University Hospital Bonn, Bonn, Germany*

MICHAEL BEILHARZ • *Institute of Innate Immunity, University Hospitals, University of Bonn, Bonn, Germany*

OLIVIA BELBIN • *Memory Unit, Neurology Department, Hospital de la Santa Creu i Sant Pau, Barcelona, Spain; Centro de Investigación Biomédica en Red sobre Enfermedades Neurodegenerativas (CIBERned), Madrid, Spain*

DAVID BROUGH • *Faculty of Life Sciences, University of Manchester, Manchester, UK*

SONIA CARTA • *Cell Biology Unit, IRCCS Azienda Ospedaliera Universitaria San Martino-IST, Genoa, Italy*

MÒNICA COMALADA • *Institute for Research in Biomedicine (IRB Barcelona), Barcelona, Spain*

VINCENT COMPAN • *Institut de Génomique Fonctionnelle, Labex ICST, Centre National de la Recherche Scientifique, Unité Mixte de Recherche 5203, Université Montpellier, Montpellier, France; Institut National de la Santé et de la Recherche Médicale Unité 1191, Montpellier, France*

ISABELLE COUILLIN • *INEM, CNRS, UMR7355, University of Orleans, Orleans, France; Molecular and Experimental Immunology and Neurogenetics, Orleans, France*

CATHERINE DIAMOND • *Faculty of Life Sciences, University of Manchester, Manchester, UK; Singapore Immunology Network (SIgN), Agency for Science Technology and Research (A*STAR), Singapore, Singapore*

ANDREA DORFLEUTNER • *Division of Rheumatology, Department of Medicine, Feinberg School of Medicine, Northwestern University, Chicago, IL, USA*

SIMONETTA FALZONI • *Department of Morphology, Surgery and Experimental Medicine, University of Ferrara, Ferrara, Italy*

BÁRBARA FLIX • *Memory Unit, Neurology Department, Hospital de la Santa Creu i Sant Pau, Barcelona, Spain*

BERNARDO S. FRANKLIN • *Institute of Innate Immunity, University Hospitals, University of Bonn, Bonn, Germany*

LORENZO GALLUZZI • *Gustave Roussy Cancer Campus, Villejuif, France; INSERM, U1138, Paris, France; 'Equipe 11 labellisée par la Ligue Nationale contre le Cancer, Centre de Recherche des Cordeliers, Paris, France; Université Paris Descartes/Paris V, Sorbonne Paris Cité, Paris, France; Université Pierre et Marie Curie/Paris VI, Paris, France*

MARCO GATTORNO • *UO Pediatria 2, Istituto G. Gaslini, Genoa, Italy*

AURÉLIE GOMBAULT • *INEM, CNRS, UMR7355, University of Orleans, Orleans, France*

BAPTISTE GUEY • *INSERM U1052, Centre de Recherche en Cancérologie de Lyon, Lyon, France; CNRS UMR5286, Centre de Recherche en Cancérologie de Lyon, Lyon, France; Université de Lyon, Lyon, France; Centre Léon Bérard, Lyon, France*

VEIT HORNUNG • *Institute of Molecular Medicine, University Hospital, University of Bonn, Bonn, Germany; Gene Center and Department of Biochemistry, Ludwig-Maximilians-University Munich, Munich, Germany*

YVAN JAMILLOUX • *Department of Biochemistry, University of Lausanne, Epalinges, Switzerland; International Research Center on Infectiology (CIRI), Inserm U1111, CNRS UMR5308 – ENS de Lyon, Lyon, France; Department of Internal Medicine, Hopital de la Croix-Rousse, Université Claude Bernard-Lyon 1, Lyon, France*

ANN M. JANOWSKI • *Graduate Program in Immunology, University of Iowa Carver College of Medicine, Iowa City, IA, USA*

MARIA KAMPES • *Institute of Molecular Medicine, University Hospital, University of Bonn, Bonn, Germany*

SONAL KHARE • *Division of Rheumatology, Department of Medicine, Feinberg School of Medicine, Northwestern University, Chicago, IL, USA*

GUIDO KROEMER • *INSERM, U1138, Paris, France; Equipe 11, Centre de Recherche des Cordeliers, Paris, France; Université Paris Descartes/Paris V, Sorbonne Paris Cité, Paris, France; Pôle de Biologie, Hôpital Européen Georges Pompidou, AP-HP, Paris, France; Metabolomics and Cell Biology Platforms, Gustave Roussy Cancer Campus, Villejuif, France; Université Pierre et Marie Curie/Paris VI, Paris, France*

MING-ZONG LAI • *Institute of Molecular Biology, Academia Sinica, Taipei, Taiwan; Institute of Immunology, National Taiwan University, Taipei, Taiwan*

EICKE LATZ • *Institute of Innate Immunity, University Hospitals, University of Bonn, Bonn, Germany; German Center for Neurodegenerative Diseases, Bonn, Germany; Centre of Molecular Inflammation Research, Department of Cancer Research and Molecular Medicine, Norwegian University of Science and Technology, Trondheim, Norway*

ED C. LAVELLE • *Adjuvant Research Group, School of Biochemistry and Immunology, Trinity Biomedical Sciences Institute, Trinity College, Dublin, Ireland*

GLORIA LÓPEZ-CASTEJÓN • *Manchester Collaborative Centre of Inflammation Research, Faculty of Life Sciences, The University of Manchester, Manchester, UK*

GWENOLA MANIC • *Regina Elena National Cancer Institute, Rome, Italy*

FABIO MARTINON • *Department of Biochemistry, University of Lausanne, Epalinges, Switzerland*

FÁTIMA MARTÍN-SÁNCHEZ • *Inflammation and Experimental Surgery Unit, Institute for Bio-Health Research of Murcia (IMIB-Arrixaca), Clinical University Hospital Virgen de la Arrixaca, El Palmar, Murcia, Spain*

ALESSANDRA MORTELLARO • *Singapore Immunology Network (SIgN), Agency for Science Technology and Research (A*STAR), Singapore, Singapore*

NATALIA MUÑOZ-WOLF • *Adjuvant Research Group, School of Biochemistry and Immunology, Trinity Biomedical Sciences Institute, Trinity College, Dublin, Ireland*

DOMINIC DE NARDO • *Institute of Innate Immunity, University Hospitals, University of Bonn, Bonn, Germany; Inflammation Division, Walter and Eliza Hall Institute (WEHI), Parkville, VIC, Australia; Department of Medical Biology, The University of Melbourne, Parkville, VIC, Australia*

PABLO PALAZÓN-RIQUELME • *Manchester Collaborative Centre of Inflammation Research, Faculty of Life Sciences, The University of Manchester, Manchester, UK*

PAWEL PASZEK • *Faculty of Life Sciences, University of Manchester, Manchester, UK*

PABLO PELEGRÍN • *Molecular Inflammation Group, Murcia Biomedical Research Institute (IMIB-Arrixaca), Hospital Virgen de la Arrixaca, Murcia, Spain*

VIRGINIE PETRILLI • *INSERM U1052, Centre de Recherche en Cancérologie de Lyon, Lyon, France; CNRS UMR5286, Centre de Recherche en Cancérologie de Lyon, Lyon, France; Université de Lyon, Lyon, France; Centre Léon Bérard, Lyon, France*

PAOLO PINTON • *Department of Morphology, Surgery and Experimental Medicine, University of Ferrara, Ferrara, Italy*

ALEXANDER D. RADIAN • *Division of Rheumatology, Department of Medicine, Feinberg School of Medicine, Northwestern University, Chicago, IL, USA*

NICOLAS RITEAU • *Immunobiology Section, Laboratory of Parasitic Diseases, National Institute of Allergy and Infectious Diseases, National Institutes of Health, Bethesda, MD, USA*

ANNA RUBARTELLI • *Cell Biology Unit, IRCCS Azienda Ospedaliera Universitaria San Martino – IST, Genoa, Italy*

VALENTINA SICA • *Gustave Roussy Cancer Campus, Villejuif, France; INSERM, U1138, Paris, France; Equipe 11, Centre de Recherche des Cordeliers, Paris, France; Faculté de Medicine, Université Paris Sud/Paris XI, Le Kremlin-Bicêtre, France; Université Pierre et Marie Curie/Paris VI, Paris, France*

DAVID G. SPILLER • *Faculty of Life Sciences, University of Manchester, Manchester, UK*

CHRISTIAN STEHLIK • *Division of Rheumatology, Department of Medicine, Feinberg School of Medicine, Northwestern University, Chicago, IL, USA*

FAYYAZ S. SUTTERWALA • *Graduate Program in Immunology, University of Iowa Carver College of Medicine, Iowa City, IA, USA; Inflammation Program, Department of Internal Medicine, University of Iowa Carver College of Medicine, Iowa City, IA, USA; Veterans Affairs Medical Center, Iowa City, IA, USA*

FRANCESCO DI VIRGILIO • *Department of Morphology, Surgery and Experimental Medicine, University of Ferrara, Ferrara, Italy*

ILIO VITALE • *Regina Elena National Cancer Institute, Rome, Italy; Department of Biology, University of Rome "Tor Vergata", Rome, Italy*

MICHAEL R. WHITE • *Faculty of Life Sciences, University of Manchester, Manchester, UK*

YUNG-HSUAN WU • *Institute of Molecular Biology, Academia Sinica, Taipei, Taiwan*

CATRIN YOUSSIF • *Institute for Research in Biomedicine (IRB Barcelona), Barcelona, Spain*

Chapter 1

Innate Immune Receptors

Natalia Muñoz-Wolf and Ed C. Lavelle

Abstract

For many years innate immunity was regarded as a relatively nonspecific set of mechanisms serving as a first line of defence to contain infections while the more refined adaptive immune response was developing. The discovery of pattern recognition receptors (PRRs) revolutionised the prevailing view of innate immunity, revealing its intimate connection with adaptive immunity and generation of effector and memory T- and B-cell responses. Among the PRRs, families of Toll-like receptors (TLRs), C-type lectin receptors (CLR), retinoic acid-inducible gene-I (RIG-I)-like receptors (RLRs) and nucleotide-binding domain, leucine-rich repeat-containing protein receptors (NLRs), along with a number of cytosolic DNA sensors and the family of absent in melanoma (AIM)-like receptors (ALRs), have been characterised. NLR sensors have been a particular focus of attention, and some NLRs have emerged as key orchestrators of the inflammatory response through the formation of large multiprotein complexes termed inflammasomes. However, several other functions not related to inflammasomes have also been described for NLRs. This chapter introduces the different families of PRRs, their signalling pathways, cross-regulation and their roles in immunosurveillance. The structure and function of NLRs is also discussed with particular focus on the non-inflammasome NLRs.

Key words Innate receptors, Toll-like receptors, NOD-like receptors, MyD88, Pattern-recognition receptors, Non-inflammasome NLRs

1 PRRs, Ancient Receptors and the Answer to a One Hundred-Year-Question

The host response to invading pathogens is an essential physiological response; hence, maintenance of an organism's integrity in the face of such challenges has been a driving force in evolution. Indeed evidence for a "defence system" can be traced back to prokaryotes [1].

Before the molecular era in immunology, the notion that the immune system had evolved to defend the host from invaders was already accepted. However, it took almost one century to identify the mechanisms underlying immune recognition. In 1884, Élie Metchnikoff observed that cells of the water flea Daphnia could engulf and destroy spores of a yeast-like fungus with "some sort of secretion". He named these cells phagocytes [2] and for the first time described three functions that we now recognise as key attributes of the innate immune system: swift detection of microbes,

Francesco Di Virgilio and Pablo Pelegrín (eds.), *NLR Proteins: Methods and Protocols*, Methods in Molecular Biology, vol. 1417, DOI 10.1007/978-1-4939-3566-6_1, © Springer Science+Business Media New York 2016

phagocytosis and antimicrobial activity. During the twentieth century, the research of Paul Ehrlich and later of Karl Landsteiner shifted the focus of attention from phagocytes to humoral immunity. In the early 1900s, Ehrlich proposed his "side-chain theory", anticipating the existence of a mechanism of immune recognition based on what was later described as the antigen–antibody interaction [3]. Then in 1933, Karl Landsteiner characterised the specificity of the antibody-antigen interaction opening the molecular era of immunology [4]. "The clonal selection theory of acquired immunity" was introduced by Frank M. Burnet. Burnet's theory explained how the specificity of antibodies was generated in the first place [5], becoming a central paradigm in immunology for nearly 50 years, bringing adaptive immunity to the centre of attention of the scientific community.

For many years adaptive immunity was the subject of intense research and the remarkable diversity of the adaptive receptors overshadowed innate immunity. The immune response was conceived as a two-compartment system in which the early innate response was seen as an unsophisticated array of mechanisms containing the infection, while the more complex adaptive response was being generated to finally eliminate the pathogen and give rise to immunological memory. However, this paradigm was unable to explain a very basic observation: how primitive organisms lacking the adaptive components were able to protect themselves and distinguish self from non-self?

Even though diversification of living creatures has led to a multiplicity of non-self recognition strategies, the key molecular principles of discrimination seem to be conserved among phyla [1]. These observations led to the idea that the templates for innate immunity have been conserved from primitive life forms to humans and that discrimination of self vs non-self and recognition of pathogens rely on phylogenetically ancient first-line sensors that recognise invariant non-self patterns. Clearly not all the encounters taking place within the course of a life cycle will pose a threat to the host. Hence, the onset of the immune response must also be tightly regulated and directed to specific targets that may put the host's integrity at risk and to avoid self-recognition. This implies that the recognition of the invader must precede the onset of any effector mechanism and also contribute to instruct the system to mount an appropriate response.

The contemporary view of the innate immune system was revolutionised in 1989 by Charles A. Janeway Jr.. In his monograph "Approaching the Asymptote? Evolution and Revolution in Immunology", Janeway Jr. introduced the concept of "pattern recognition receptors" (PRRs) [6] postulating that PRRs recognising microbial-derived products link innate and adaptive immunity by activating antigen-presenting cells (APCs) to provide the second signal required for T-cell activation and initiation of the adaptive response. He proposed that, opposed to the adaptive receptors, PRRs were non-clonally distributed receptors encoded by single non-rearranging genes. The proposed function of PRRs was to recognise structural patterns in molecules found in microorganisms but not in multicellular organisms to efficiently differentiate noninfectious self from infectious non-self. Janeway postulated that these conserved "pathogen-associated molecular patterns" (PAMPs, now also referred to as MAMPs for "microbe-associated molecular patterns") recognised by PRRs should be the result of a specific metabolic pathway characteristic for the microorganisms like a carbohydrate or lipopolysaccharide absent from the host.

He also reasoned that the adaptive immune response required two signals for activation: ligation of the specific receptor on the surface of a T or B cell by the antigen and a second signal derived from the antigen-presenting cell later identified as costimulatory molecules [6]. Janeway's lab subsequently established that several components of bacteria, yeast, and viruses had the ability to enhance costimulatory activities for T cells [7] and also demonstrated that cis-presentation of both antigen and costimulators was needed for T-cell activation [8] linking innate and adaptive immunity.

Later on, Polly Matzinger challenged Janeway's theory introducing the "danger theory", suggesting that the main determinant of immune activation is not the origin of the antigen itself but the extent of damage. Consequently, instead of sensing PAMPs or MAMPs, the immune system would recognise danger-associated molecular patterns (DAMPs) that could be produced by the host itself [9]. Anything that can cause tissue damage, whether of microbial or nonmicrobial origin, can be sensed and will trigger an immune response, while if the stimulus does not pose any hazard to the host, even being a microbe, it will be "ignored".

While these theories introduced new concepts on how innate recognition contributes to self–non-self discrimination, the discovery of the first PRR, a member of the Toll-like receptors (TLRs) [10, 11], provided essential proof. The discovery of TLRs subsequently led to the characterisation of other families of innate immune receptors and their ligands, and later studies revealed DAMPs, including ATP, heparan sulphate, HMGB1 and S-100 proteins that can trigger immune responses upon ligation of innate receptors [12].

In addition to this essential role in sensing microbes and damage, innate immunity also regulates and directs the activation of the adaptive immune system through polarisation of antigen-presenting cells equipped with the germline-encoded PRRs, shaping the overall outcome of the response (Fig. 1). Haematopoietic cells, including dendritic cells, macrophages, and neutrophils, and even T and B cells, as well as non-haematopoietic cells such as epithelial cells, contribute to this host-defence system by expressing different arrays of PRRs.

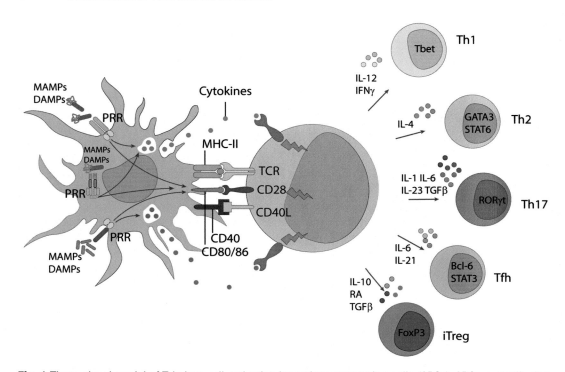

Fig. 1 Three-signal model of T-helper cell activation by antigen-presenting cells (APCs). APCs, typically dendritic cells (DC), sense microbial components (microbe-/danger-associated molecular pattern, MAMPs/DAMPs) through pattern recognition receptors (PRRs) triggering intracellular signalling cascades. This activates DCs, enhancing antigen uptake and processing for presentation in MHC class II molecules. Antigen-MHC-II complex constitutes signal 1 for the T-helper (Th) cell that interacts with it through its specific T-cell receptor (TCR). PAMP–PRR interaction also stimulates expression of costimulatory molecules on the APC, such as CD40, CD80 and CD86 that will constitute signal 2 for the Th cell. Signal 3 is given by the polarising cytokines and other various soluble or membrane-bound factors, such as interleukin (IL-) 12, interferon gamma (IFNγ), IL-4, IL-1, IL-6 IL-21, IL23, IL-10, tumour growth factor beta (TGFβ) or retinoic acid (RA). The specific combination of polarising cytokines promotes the development of Th1, Th2, Th17, T follicular helper cells (Tfh) or inducible T regulatory cells (iTreg). While the specific profile of T-cell-polarising factors is triggered by recognition of specific MAMPs and DAMPs by an array of PRRs, the interaction between CD40 on the APC and CD40-ligand (CD40L) expressed on the activated T cell contributes to stabilise the phenotype. *STAT* signal transducers and activators of transcription, *Tbet*: T-box transcription factor, *GATA*: globin transcription factor, *ROR*: RAR-related orphan receptors, *Bcl*: B-cell CLL/lymphoma 2, *FoxP3*: forkhead box P3

To date, along with TLRs, several other families of innate receptors have been characterised. PRRs can be subdivided into membrane-bound receptors that include TLRs, along with C-type lectin receptors (CLR), and cytoplasmic receptors including retinoic acid-inducible gene-I (RIG-I)-like receptors (RLRs) and the nucleotide-binding domain, leucine-rich repeat-containing protein receptors (NLRs). A number of other PRRs including the cytosolic DNA sensor cGAS (cyclic GMP-AMP synthase) and the family of absent in melanoma (AIM)-like receptors (ALRs) have also been recently described [13].

When these receptors bind their agonists, they trigger an innate immune response by engaging certain signalling cascades that ultimately activate transcription factors such as nuclear factor kappa B (NFκB), activator protein-1 (AP-1), ETS domain-containing protein Elk-1, activating transcription factor 2 (ATF2), the phosphoprotein p53 and members of the interferon-regulatory factor (IRF) family, leading to specific gene expression programmes. Several of the genes being expressed encode chemokines, such as interleukin (IL)-8, CCL-2 and CXCL-1 that promote recruitment of leukocytes including neutrophils, monocytes and lymphocytes and a vast array of cytokines that will amplify the inflammatory response, enhance antigen presentation and costimulatory molecule expression, initiate tissue repair and direct T-cell polarisation and differentiation into different lineages of effector T cells (Th-1, Th-2, Th-17, regulatory T-cells (Tregs), among others) [14].

2   Toll-Like Receptors

The Toll-like receptors are the prototypical innate pattern recognition receptors that sense danger- and microbial-associated molecular patterns.

The first clues that linked TLRs to innate immunity came from studies carried out in the fruit fly *Drosophila melanogaster*. The founding member of the TLRs, the Toll protein, was initially identified as a gene product essential for the development of embryonic dorsoventral polarity in the fly [15]. Later, the protein Toll was shown to share homology with the previously identified interleukin-1 receptor 1 (IL-1R1) [16] through which the pleiotropic pro-inflammatory cytokine IL-1 exerts its effects [17]. The first striking finding was that, even though both proteins had dissimilar physiological functions, they contained similar amino acid sequences known to be essential for NFκB signalling [18], a factor originally described to mediate the response to lipopolysaccharide in B cells [19]. Finally, in 1996 the work of Bruno Lemaitre showed the involvement of the protein Toll in the antifungal response in *D. melanogaster* and production of the antifungal peptide drosomycin, confirming its role in innate immunity [10].

In 1997 the first human homolog for the *Drosophila* Toll protein was described by R. Medzhitov in Janeway's lab [11]. To date, 13 members of the TLR family have been identified in mammals including 10 human TLRs (TLR1–TLR10) and 12 murine TLRs (TLR1–TLR9 and TLR11–TLR13). Although most of the TLRs are conserved between humans and mice, TLR10 has lost its functionality in mice due to a retroviral insertion; TLR11, TLR12 and TLR13 are missing in the human genome [20]. Orthologs and paralogs for several mammalian TLRs have been also identified in different taxa including birds, amphibians, teleosts and agnathans.

In addition to insects, TLRs have been also traced back to ancient invertebrates including sponges, cnidarians, oligochaetes, molluscs and crustaceans [21].

2.1 Structure and Ligand Recognition in TLRs

Biochemically, TLRs are defined as a family of type-I transmembrane glycoproteins, typically composed of three domains: the N-terminal ectodomains, characterised by the presence of leucine-rich repeat (LRR) motifs which dictate ligand specificity, either by direct interaction or through accessory molecules, a hydrophobic transmembrane domain and the internal C-terminal domain that mediates intracellular signalling [22].

TLRs can be found either inserted in the cellular membrane or as membrane-bound proteins in endosomes. The Toll-like receptors 1, 2, 4, 5 and 6 are found primarily, but not exclusively, in the plasma membrane; conversely, TLR3, TLR7, TLR8, TLR9 and the murine TLR11, TLR12 and TLR13 are localised in intracellular endosomal and endolysosomal compartments [13]. Trafficking of TLRs is a tightly regulated process, and endosomal localisation normally requires UNC93B1, a transmembrane protein known to control the movement of TLRs from the endoplasmic reticulum where the assembly of TLRs takes place, to their final location in endosomes [23, 24].

The LRR portion of the TLR is responsible for ligand specificity [25, 26]. These ectodomains recognise a wide variety of biomolecules that can be derived from bacteria, fungi and parasites or endogenously generated (Table 1). The LRR is either extracellular or facing the luminal compartment of endosomes where they encounter molecules released by invading pathogens or damaged tissue. Typically they present a horseshoe form as described for other LRR-containing proteins [22]. However, the proposed crystallographic structures for several TLR–ligand complexes have revealed that, in contrast to what has been observed for most of LRR-containing proteins, ligand binding to the LRR portion of the TLRs occurs most often on the ascending lateral surface of the ectodomains [25, 27–29].

Comparative sequence analysis of the vertebrate LRRs grouped TLRs into six subfamilies, revealing that TLRs from different species grouped according the primary sequence of their ectodomains recognising similar types of ligands. This suggested that selective pressure to maintain specificity for certain ligands has dominated the evolution of the ectodomains. Among the mammalian subfamilies, the TLR1 subfamily containing TLR1, TLR2 and TLR6 is associated with recognition of lipoproteins and lipopeptides; the TLR3 subfamily recognises double-stranded RNA; the TLR4 subfamily is linked to recognition of lipopolysaccharides; the TLR5 subfamily recognises the structural protein of the bacterial flagellum, flagellin; and the TLR7 subfamily comprising TLR7–TLR9 recognises nucleic acids [30]. The TLR11 subfamily including

Table 1
Mouse and human TLR expression, ligands and pathogens recognised

TLR	Localisation	Species-specific expression	Natural ligands	Synthetic ligands	Type of pathogen recognised
TLR1	Extracellular	Human/mice	Triacyl lipopeptides	Pam3CSK4	Bacteria
TLR2	Extracellular	Human/mice	Lipoproteins, peptidoglycan, LTA, zymosan/mannan	Pam3CSK4	Bacteria
TLR3	Endosomal	Human/mice	dsRNA	polyI:C and polyU	Viruses
TLR4	Extracellular/ Endosomal	Human/mice	LPS, RSV, mannans and glycoinositolphosphate from *Trypanosoma* spp.	Lipid A derivatives	Gram-negative bacteria and viruses
TLR5	Extracellular	Human/mice	Flagellin	ND	Bacteria
TLR6	Extracellular	Human/mice	Diacylipopeptides LTA and zymosan	MALP2	Bacteria
TLR7	Endosomal	Human/mice	ssRNA/short dsRNA	Imidazoquinolines and guanosine analogues	Viruses and bacteria
TLR8	Endosomal	Human/mice	ssRNA/short dsRNA	Imidazoquinolines and guanosine analogues	Viruses and bacteria
TLR9	Endosomal	Human/mice	CpG DNA hemozoin *Plasmodium* spp.	CpG ODNs	Bacteria, viruses and protozoan parasites
TLR10	ND	Human	ND	ND	ND
TLR11	Endosomal	Mice	Profilin/flagellin	ND	Apicomplexan parasites and bacteria
TLR12	Endosomal	Mice	Profilin	ND	Apicomplexan parasites
TLR13	Endosomal	Mice	Bacterial 23S rRNA	ND	Gram-negative and Gram-positive bacteria

ND not defined, *dsRNA* double-stranded RNA, *LPS* lipopolysaccharide, *LTA* lipoteichoic acid, *MALP2* macrophage-activating lipopeptide 2, *ODN* oligodeoxynucleotide, *polyI:C* polyinosinic–polycytidylic acid, *polyU* poly-uridine, *rRNA* ribosomal RNA, *RSV* respiratory syncytial virus, *ssRNA* single-stranded RNA, *TLR* Toll-like receptor

murine TLR11, TLR12 and TLR13 has been the least explored so far, probably because these receptors are absent in humans [31]. Their natural ligands have been identified only recently, revealing that similar to TLR5, TLR11 and TLR12 recognise proteins. Originally TLR11 and TLR12 were reported to recognise profilin, a protein derived from apicomplexan parasites like *Toxoplasma gondii* [32, 33]. Surprisingly a recent study reported that flagellin, previously reported as a ligand for TLR5, is also a ligand for TLR11 [34, 35]. Likewise, new studies revealed that TLR13 acts as a receptor for bacterial ribosomal RNA 23S [36].

Upon ligand binding, TLRs undergo a molecular rearrangement leading to the two extracellular domains forming an "m"-shaped homo- or heterodimer with the ligand staying in between the two receptors in a "sandwich-like" arrangement. This conformational change brings the transmembrane and cytoplasmic domains into close proximity, allowing the C-terminal TIR domains to generate an active interacting domain that triggers the intracellular signalling cascade. Recent studies have shown that the transmembrane domain (TMD) regions have a pivotal role during receptor oligomerisation. Strikingly, it was shown that isolated TMDs lacking the ectodomains and intracellular TIR domains replicate the homotypic and heterotypic interactions with the same partner receptors as the full length proteins, revealing the importance of this region for the interaction between TLRs [37].

The cytoplasmic signalling C-terminal domain presents homology to the IL-1R and is thus referred as the Toll-IL-1-resistance (TIR) domain. The TIR domain of the TLRs interacts with TIR-domain-containing adaptor molecules in the cytosol which in turn trigger downstream signalling pathways that lead to the expression of proinflammatory cytokines, chemokines, antiviral and antibacterial proteins, among others [38].

Notwithstanding the substantial progress on the structural characterisation of TLRs, more information is required to fully understand the interaction between each TLR and its proposed ligand. There are still no crystal structures available for several TLRs including mammalian TLR5 and TLRs 7–13. If ligand recognition is mediated by yet uncharacterised proteins bridging the interaction between the ligand and the LRRs, as is the case for TLR4 and LPS interaction, they will need to be elucidated. Finally, more information is needed to understand how TLR-TIR domains interact with each other or with the TIRs of adaptor molecules.

2.2 TLR Signalling Pathways

Typically, upon ligand recognition TLRs experience conformational changes that are critical for the recruitment of TIR-domain-containing proteins to the TIR domain of the receptor and transduction of the signal. There are five TIR-domain adaptor molecules: myeloid differentiation primary-response protein 88 (MyD88), MyD88-adaptor-like (MAL) also known as

TIR-associated protein (TIRAP), TIR-domain-containing adaptor protein-inducing IFN-β (TRIF) also named TIR-domain-containing molecule 1 (TICAM1) and TRIF-related adaptor molecule (TRAM) and sterile-α-and armadillo-motif-containing protein 1 (SARM1) [39].

MyD88 and TRIF act as switches for distinct signalling pathways that in turn activate two important families of transcription factors involved in regulation of several genes that are implicated in the control of the immune response. The MyD88 pathway ultimately, but not exclusively, leads to the nuclear translocation of the transcription factor NFκB, whereas the TRIF pathway mainly triggers translocation of the IRFs, particularly IRF3.

NFκB proteins regulate expression of a diverse array of genes involved in control of innate and adaptive immunity, cell cycle, anti-apoptotic response and stress responses. In the context of innate responses, NFκB has been implicated in the induction of genes encoding proinflammatory cytokines and leukocyte recruitment [40].

The family of IRF transcription factors plays important roles in cell growth, survival and differentiation of haematopoietic cells, a key function being the orchestration of antiviral responses through the induction of type-I interferons (IFN-I) [41].

In the past decade, several studies suggested that IRFs can also be activated in a MyD88-dependent fashion, and it is now widely accepted that the MyD88-IRF axis makes a major contribution to the immune response triggered by TLR activation. In the next sections MyD88-dependent and MyD88-independent signalling pathways are introduced followed by an overview of the role of IRFs in TLR signalling.

2.2.1 MyD88 Signalling Pathway

MyD88 is recruited by all TLRs except for TLR3, upon ligand recognition (Fig. 2). The first event in the MyD88 signalling pathway is the formation of a complex involving IL-1R-associated kinase (IRAK) members and MyD88 adaptor named the "myddosome" [42]. MyD88 associated with the cytoplasmic portion of TLRs interacts with IRAK members through homophilic interactions of the death domains. IRAK members associate with TRAF6, which in turn activates transforming growth factor-activated kinase 1 (TAK1). TAK1 then activates the I kappa B kinase (IKK) complex and mitogen-activated protein kinase (MAPK) pathway [43].

The IKK complex is the core element of the NFκB cascade, and it is essentially composed of two kinase subunits, IKKα and IKKβ, and a regulatory subunit, NEMO/IKKγ. NFκB is a family of transcription factors that while inactive, is kept in the cytosol through interaction with members of the IκB family. The TAK1 complex activates the IKK complex by phosphorylation, which in turn phosphorylates IκB proteins, allowing their ubiquitination and degradation by the proteasome. IκB degradation releases NFκB,

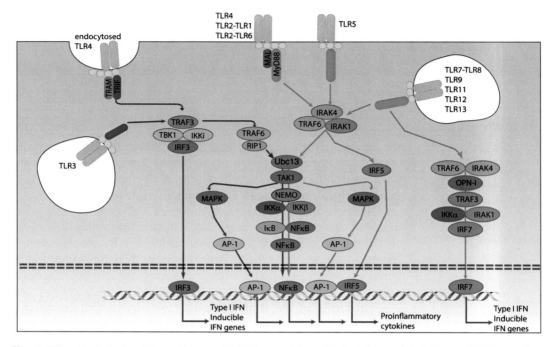

Fig. 2 TLR-activated signalling pathways. MyD88 associates with IL-1R-associated kinase (IRAK) members forming the myddosome. IRAK4 activates IRAK1, which in turn catalyses its autophosphorylation before it is released from the myddosome. IRAK1 then associates with TRAF6, an E3 ligase that together with UBC13 and UEV1A catalyses its own ubiquitination as well as ubiquitination of the transforming growth factor-activated kinase 1 (TAK1) protein complex formed by TAB1, TAB2 and TAB3. TAK1 then activates the I kappa B kinase (IKK) complex and mitogen-activated protein kinase (MAPK) pathway. The TAK1 complex activates the IKK complex by phosphorylating the IKKβ subunit. In turn, the active IKK complex phosphorylates IκB proteins which allows ubiquitination and degradation by the proteasome. As a result of IκB degradation, NFκB is released. Free NFκB translocates into the nucleus. Effector kinases of the MAPK pathway JNK, p38 and ERK are also activated, leading to AP-1 translocation into the nucleus to activate transcription of inflammatory genes. IRF5 can be also recruited to the MyD88–IRAK4–TRAF6 complex, phosphorylated and translocated to the nucleus to promote expression of proinflammatory cytokines. TLR4, TLR1/TLR2 and TLR2/TLR6 require recruitment of the adaptor MAL to activating the MyD88-dependent pathway. TRIF is recruited to TLR3 and endosomal TLR4. Endosomal TLR4 also requires recruitment of TRAM to initiate signalling. TRAF3 activates TBK1 and IKKi, which mediate phosphorylation of IRF3 triggering its dimerisation. IRF dimers translocate to the nucleus to induce expression of type-I IFN and IFN-inducible genes. TRIF also interacts with TRAF6 and RIP1, mediating NFκB activation. Endosomal TLRs sensing nucleic acids can activate the MyD88–TRAF6–IRF7 axis. Preferentially in plasmacytoid DCs, a complex consisting of MyD88–TRAF6–IRAK4–IRAK1–IRF7 is formed. OPN-i, TRAF3 and IKKα are also involved in this complex. Formation of the complex triggers IRF7 phosphorylation by IRAK1 and subsequent translocation to the nucleus to induce expression of type-I IFN and IFN-inducible genes

consisting of p65 (also known as RelA), c-Rel and p50 which translocates into the nucleus to activate transcription of cytokine genes associated with inflammation including TNF-α, IL-1β, IL-6 and IL-12p40; genes encoding cell adhesion and recruitment molecules like CXC and CC chemokines; and growth factors and antiapoptotic signals [44].

MAPK pathway activation results in the activation of the effector kinases c-Jun N-terminal kinase (JNK), p38 and extracellular signal-regulated kinase (ERK). Following TLR dimer formation in response to ligand recognition, activation of TRAF6 and TAK1 will activate the kinases MKK3 and MKK6, which will phosphorylate p38 and JNK, respectively. JNK phosphorylates c-Jun which binds to c-Fos to form the complex known as AP-1, which is then translocated into the nucleus to activate transcription of inflammatory genes [45].

Additionally, the p38 pathway can regulate gene expression through phosphorylation of the transcription factor cAMP response element-binding protein (CREB). TLR-dependent phosphorylation of CREB enhances its transactivation potential and plays an important role in regulating the transcriptional induction of many proinflammatory mediators, including cyclooxygenase 2 (COX-2) and TNF-α [46].

The role of the ERK pathway in TLR-induced responses has received less attention, but it is known to regulate gene expression at transcriptional and posttranscriptional levels. Another MAP3K known as Tpl2 is used instead of TAK1 to activate the ERK pathway downstream most of the TLRs [47].

The generation of MyD88 knockout mice and cell lines confirmed the crucial role of MyD88 in proinflammatory cytokine production and NFκB activation upon TLR ligation. Deficiency in the MyD88 signalling pathway resulted in impaired inflammatory cytokine secretion in response to several TLR agonists including the ligand for TLR2/TLR6, mycoplasmal macrophage-activating lipopeptide-2 (MALP-2) [48], CpG DNA which signals through TLR9 [49], the TLR5 ligand flagellin [50] and LPS and the ligand for TLR4 [51].

MyD88 deficiency also impaired cytokine secretion and NFκB activation in the response to IL-1β and IL-18, but not to TNF-α, IL-2 or IL-4 [51, 52]. While TNF-α, IL-2 and IL-4 signal through unrelated receptors, IL-1β, IL-18 and all the TLR ligands require different receptors of the TLR–IL-1R superfamily [53]. These observations suggested that MyD88 is a universal adaptor for this receptor superfamily. However, not all the effects induced by the TLR4 ligand LPS or the TLR3 ligand poly(I:C) were completely abrogated in MyD88 knockout mice, pointing to the existence of an alternative MyD88-independent signalling pathway that was exclusively activated upon engagement of TLR4 and TLR3.

2.2.2   TRIF-Dependent Pathway

Several observations contributed to the hypothesis that a MyD88-independent TLR signalling pathway existed. First, it was reported that LPS and poly(I:C) were able to induce dendritic cell maturation in MyD88$^{-/-}$ dendritic cells as revealed by upregulation of costimulatory molecules [51, 54]. Also, even though degradation of IκBα was delayed in MyD88$^{-/-}$ macrophages stimulated with poly(I:C) or LPS, JNK and p38 were activated to a similar extent and with comparable kinetics to those seen in the wild-type cells [54]. Besides, despite the fact these cells failed to produce inflammatory cytokines, NFκB and MAPK were activated, albeit with delayed kinetics [55]. Moreover, these cells responded to TLR4 and TLR3 agonists by secreting IFN-β [56] and IP-10 (CXCL-10) [56]. TRIF was identified as the alternative adaptor molecule downstream of TLR4 and TLR3 [57–59]. Generation of TRIF knockout mice confirmed that TRIF was required for IFN-β production in cells stimulated with LPS or poly(I:C) and late activation of NFκB and MAPK was abolished in MyD88/TRIF knockout mice [58].

Interestingly, TRIF has the ability to trigger both IRF and NFκB translocation to the nucleus, activating type-I interferon (IFN-I) and interferon-inducible genes as well as transcription of inflammatory genes (Fig. 2). Amino and carboxy-terminal domains of TRIF have a different ability to bind proteins that act downstream in the signalling pathway. The C-terminal region interacts with receptor-interacting protein 1 (RIP1) kinase through its RHIM (RIP homotypic interaction motif), which after ubiquitination forms a complex with TRAF6 and TAK1. Ultimately, the formation of this complex will activate TAK1 and result in NFκB and MAPK activation, but not IFN-β secretion [60]. On the other hand, the N-terminal domain recruits the noncanonical IKKs TBK1 and IKKi and TRAF3, leading to activation of IRF3, which after forming a dimer translocates into the nucleus to induce transcription of IFN-I genes including IFN-β. The N-terminal domain can also recruit TRAF6, leading to nuclear translocation of NFκB and proinflammatory cytokine secretion [60].

TLR4 is unique in its capacity to activate both the MyD88 and TRIF pathways and entails the most complex signalling machinery of all TLRs. TLR4 uses the adaptor protein MAL as a bridge between the receptor and MyD88. MAL is also used by TLR2 although to a lesser extent. The adaptor TRAM links TRIF to TLR4 to induce IRF3 signalling. Subcellular localisation seems to be a critical factor for the activation of the TRIF pathway; indeed all TLRs activating this pathway are localised in endosomes. For the particular case of TLR4 upon activation, the receptor is endocytosed in endosomes. Change in its subcellular localisation acts as a switch between the MAL/MyD88 and the TRAM/TRIF signalling pathways, wich are activated sequentially rather than simultaneoulsly. [39]. SARM has been shown to be another important TIR-adaptor protein involved in the regulation of the TRIF pathway. However, in contrast to the other TIR-adaptor molecules, SARM acts as a negative regulator of TRIF [61].

2.2.3 IRF and Myd88

Apart from IRF3, other members of the IRF family of transcription factors also play important roles in MyD88-dependent signalling upon recognition of viral products through TLRs.

IRF7 has been described as a master regulator and is activated downstream of MyD88 in response to TLR7 and TLR9 ligation to induce IFN-I secretion [62]. In particular, plasmacytoid dendritic cells (pDCs) constitutively express IRF7 and respond swiftly by secreting IFN-I when exposed to viral products. The MyD88-IRF7 pathway is absolutely required for IFN-I secretion in pDCs [63]. MyD88 can directly associate with IRF7 which, when inactive, stays in the cytosol. IRF7 is subsequently phosphorylated and activated to form part of a complex composed of MyD88, IRAK1, IRAK4, TRAF3, TRAF6 and IKKα. The ubiquitin–ligase activity of TRAF6 is required for maximal activation of IRF7 [64]. The production of IFN-α in response to TLR9 ligands in pDCs requires activation of the phosphoinositide 3-OH kinase (PI3K)/mammalian target of rapamycin (mTOR) pathway [63]. The intracellular phosphoprotein osteopontin (Opn-i) that has been described as essential for the development of T-helper 1 responses also plays a key role in MyD88-IRF7 pathway in pDCs stimulated with CpG and has been found as a component of the MyD88 signal transduction complex [65]. While MyD88, IRAK4, TRAF6 and IKKα are required for NFκB and IRF7 activation, IRAK1, TRAF3 and Opn-i selectively induce activation of IRF7 [66] (Fig. 2).

Like IRF7, IRF8 also interacts with MyD88 and mediates production of IFN-I and other inflammatory cytokines when activated by TLR9 engagement. IRF8 is a nuclear protein expressed in pDCs and also in conventional dendritic cells (cDCs) [67]. It has been implicated in TLR9-induced production of IFN-I and proinflammatory cytokines and also in the amplification phase of IFN-I production during viral infections [68].

IRF5 was essential for the MyD88-dependent production of IL-6 and IL-12 in TLR-mediated responses but was not required for IFN-α production [69].

IRF1 is induced by IFN-γ and also interacts with MyD88 upon TLR activation. MyD88-IRF1 interaction induces efficient translocation of IRF1 into the nucleus. The importance of IRF1 downstream of TLR engagement is supported by studies in IRF1-deficient cells showing impaired IFN-β secretion, inducible nitric oxide synthase (iNOS) activation and IL-12p35 production in response to TLR9 or TLR3 ligands [70].

TLR activation plays a key role in promotion of both humoral and the cell-mediated immunity. Optimal TLR signalling determines the combination of cytokines that will in turn define the outcome of the adaptive immune response. These receptors work in tandem with other receptors of the innate immune system to regulate innate responses, and they are key partners of NLRs, providing the first signal that is required for assembly of inflammasomes and further amplification of inflammation.

3   C-Type Lectin Receptors

The C-type lectin receptors or CLRs comprise another important family of PRRs that play a major role in antimicrobial immunity. The CLR superfamily is divided into 17 groups (I–XVII) according to their diverse structure and phylogeny including more than 1000 proteins [71, 72].

The CLRs were first described by the presence of a calcium-dependent carbohydrate-binding motif known as the carbohydrate recognition domain (CDR). However, it was later found that there were similar structurally conserved domains able to bind diverse ligands including glycans, lipids and proteins, among others [73]. These domains are now known as C-type lectin-like domains (CTLD) and are also characteristic of the CLRs [74]. Structurally CDRs and CTLD contain a motif composed of two loops harbouring conserved cysteine residues that stabilise the structure by establishing disulphide bridges between the two chains [73]. It is now clear as well that some CLRs can bind ligands independently from Ca^{2+} [74].

Given the vast and diverse number of proteins in the CLR superfamily, general characteristics of some membrane-bound CLRs and its signalling pathways are discussed below. Detailed information on particular receptors can be found in several comprehensive reviews [71, 73, 75].

3.1   Role of CLR in Microbial Recognition

CLRs which are mainly expressed in myeloid cells can be soluble or membrane bound and sense a wide variety of self and non-self ligands [75, 76]. The membrane-bound CLRs are classified into two groups: type-I CLRs that include receptors belonging to the mannose receptor family and group II CLRs that are part of the asialoglycoprotein receptor family. The latter includes the DC-associated C-type lectin 1 (dectin 1, also known as CLEC7A) subfamily and the DC immunoreceptor (DCIR or CLEC4A) subfamily [76]. CLRs appear more promiscuous than other PRRs and have been shown to bind several types of ligands. The CLRs expressed by DCs seem to preferentially recognise mannose, fucose and glucans, which allow them to recognise most types of pathogens including bacteria, fungi, viruses and parasites. Others, including Lox-1 or DNGR-1, respond to self ligands such as dead cells, while mincle or DC-SIGN can recognise ligands of microbial and self-origin and may mediate distinct responses to each one. CLRs have also been implicated in antitumor responses [72, 77] (Table 2).

The effects of CLRs upon ligand recognition are varied. Many CLRs can promote phagocytosis and endocytosis of the ligands, leading to degradation, which favours antigen presentation to T cells. Depending on the targeted CLR, the antigen will be directed towards either the MHC class I or MHC pathway or both [76]. Some CLRs can also promote microbicidal activity in innate cells, thereby enhancing pathogen clearance [71].

Table 2

Ligands and pathogen recognised by some CLRs

CLR	Ca²⁺ requirement	Ligands	Type of pathogen recognised
Dectin-2	Ca²⁺ dependent	α-mannans O-linked mannobiose-rich glycoprotein	*M. tuberculosis, S. mansoni, S. mansoni* egg antigen, *C. albicans, Malassezia* spp.
Mincle	Ca²⁺ dependent	α-mannose mannitol-linked glyceroglycolipid mannosyl fatty acids	*M. tuberculosis, C. albicans, Malassezia* spp.
DC-SIGN	Ca²⁺ dependent	High mannose surface layer A protein	HIV-1, Measles, Dengue, *Mycobacterium* spp., Influenza A, *S. mansoni* egg antigen, *Leishmania* spp., *H. pylori, Lactobacillus* spp., *M. leprae, Bacillus Calmette Guerin*
SIGNR3	Ca²⁺ dependent	High mannose and fucose	*L. infantum* *S. mansoni* egg antigen
DCIR	Ca²⁺ dependent	Not defined	HIV
Dectin-1	Ca²⁺ independent	β-glucans	*L. infantum* *C. albicans* *Mycobacterium* spp.
Mannose receptor (MR)	Ca²⁺ dependent	High mannose mannosylated lipoarabinomannan	*S. pneumoniae* *M. corti* *Mycobacterium* spp. *K. pneumoniae* *S. pneumoniae* *F. tularensis* *S. mansoni* egg antigen
DEC-205 (CD205)	Ca²⁺ dependent	Plasminogen activator	*Y. pestis*

3.1.1 CLRs in Antifungal Immunity

A number of CLRs contribute to antifungal responses including dectin 1, dectin-2, mincle and the mannose receptor (MR). However, to date only mutations in dectin-1 have been associated with increased susceptibility to fungal infections in humans [77] suggesting that several CLRs may have redundant roles in antifungal responses.

Dectin-1 recognises β-glucan moieties present in the cell wall of fungal pathogens including *Candida, Aspergillus, Pneumocystis* and *Coccidioides* species. Activation of dectin-1 involves receptor clustering and formation of the phagocytic synapse. This is a prerequisite for intracellular signalling. Activation of the dectin-1 pathway has been shown to be critical in inducing polarisation of Th1 and Th17 cells that are essential for fighting systemic and mucosal fungal infections, respectively [77]. Besides regulating differentiation of Th cells, engagement of the dectin-1 pathway has important effects on other cellular processes including phagocytosis, respiratory burst, autophagy and production of a number of proinflammatory mediators. Importantly, dectin-1 has been shown to operate together with the NLRP3 inflammasome as well as non-canonical caspase 8 inflammasomes to induce production of IL-1β [78–80]. Additionally, dectin-1 was required for IFN-I production in the context of *Candida albicans* infections through activation of the IRF5 pathway [81].

Dectin-2 has been also implicated in antifungal responses. It recognises α-mannans from *C. albicans* and O-linked mannobiose-rich residues from the *Malassezia* (formerly *Pityrosporum*) spp. [77]. Similar to dectin 1, dectin-2 promotes Th17 responses and also stimulates production of several cytokines including IL-23 and IL-1β in addition to reactive oxygen species (ROS).

3.1.2 CLRs in Bacterial Infections

Most of the evidence implicating CLRs in antibacterial immunity came from the study of CLRs in the context of mycobacterial diseases. A number of CLRs recognise PAMPs derived from mycobacteria including mincle, dectin 1, DC-SIGN (mice SIGNR3) and dectin-2 [72]. In vitro studies implicated multiple CLRs, namely, dectin-1, DC-SIGN and MR, in *Mycobacterium tuberculosis* infection control. However, when experiments were carried out in vivo, each of these CLRs appeared to be redundant [74]. Despite the apparent redundancy of these receptors during in vivo infection, the common signalling pathway involving CARD9 seems to be critical for protection since CARD9 deficiency in mice leads to uncontrolled bacterial replication and death [82].

CLRs have been also implicated in recognition of several other bacterial pathogens. DC-SIGN can recognise *Mycobacterium leprae*, *Bacillus Calmette–Guérin* (BCG), *Lactobacilli* spp. and *Helicobacter pylori* [76].

MR can also recognise bacteria such as *Streptococcus pneumoniae* [83], *Mycobacterium kansasii* [84] and *Francisella tularensis* [85]. However, the MR seems not to be essential during infection with these pathogens in vivo [76]. DEC-205 is another example of a CLR involved in recognition of bacterial pathogens. DEC-205 can bind to plasminogen activator expressed by *Yersinia pestis*. However, instead of eliciting a protective response against the pathogen, DEC-205 was found to promote dissemination of bacteria with detrimental consequences for the host [86], suggesting that pathogens can also take advantage of the internalisation pathway offered by these receptors to evade the immune response.

3.1.3 CLRs and Viruses

The interaction between CLRs and viruses is not always beneficial for the host. Similar to the case of DEC-205 and *Y. pestis*, CLRs can favour viral infections and transmission, with detrimental consequences for the host. In this regard, DC-SIGN interaction with HIV is one of the best characterised examples. Interaction of the viral protein gp120 with DC-SIGN favours viral entry into the cells, enhancing infection of CD4+ T cells [87]. DC-SIGN has been also implicated in facilitating infection by influenza virus, which binds to the receptors through glycans on haemagglutinin [88]. Similar interactions between CLRs and other viruses such as dengue virus have been reported [76].

Despite the negative outcome that the interaction of some CLRs with viruses may have, in some cases it can be beneficial for the host and contribute to the antiviral response. CLEC9A is important for cross-presentation of antigens from vaccinia virus and *Herpes simplex* virus which is crucial for promoting cytotoxic antiviral responses [89, 90].

3.1.4 CLRs in Parasitic Invasion

CLRs have been implicated in recognition of carbohydrate moieties from parasites, particularly helminths. These parasites express a wide range of glycan moieties that can be recognised by different CLRs. DC-SIGN was implicated in the recognition of the soluble egg antigen of *Schistosoma mansoni* and other *Schistosome* spp. [91]. A range of other CLRs including MR, SIGNR1, SIGNR2 and dectin-2 have been implicated in recognition of *S. mansoni* antigens [74]. Dectin-2 has been shown to reduce Th2-mediated pathology in *S. mansoni* infection by promoting secretion of IL-1β through NLRP3 activation [92].

Finally, other infection models, particularly of central nervous system parasitic infections (e.g. neurocysticercosis), have shown that engagement of CLR by parasite ligands can contribute to pathology [74].

3.2 Signalling Downstream of CLRs

As previously mentioned, activation and signalling through CLRs has multiple outcomes including phagocytosis, activation of innate killing mechanisms by generation of microbicidal compounds such as ROS as well as production of inflammatory mediators. The immune response elicited by engagement of CLRs can be very different depending on the type of receptor, the cell-type expressing it and the nature of the ligand being recognised. Signalling pathways triggered by CLRs are only partially understood, and experimental evidence suggested that activation through CLRs such as MR, DEC-205 and cluster of differentiation (CD)-207 alone is insufficient to elicit gene transcription and/or microbicidal effector functions in myeloid cells, requiring cooperation of other receptors. The signalling pathways triggered by CLRs are complex and are often implicated in cross talk with other PRRs like TLRs and NLRs. On the other hand, some other CLRs including dectin-1, dectin-2, SIGNR3 and mincle are self-sufficient and have been shown to directly couple PAMP recognition to myeloid cell activation and adaptive immunity [93].

According to the type of cytoplasmic signalling motifs and signalling potential, CLRs expressed in the myeloid linage can be classified into different categories: Syk-coupled CLRs, immunoreceptor tyrosine-based inhibitory motif (ITIM)-expressing CLRs, CLRs without immunoreceptor tyrosine-based activation motif (ITAM) or ITIM domains [71].

3.2.1 Syk-Coupled CLR

The self-sufficient CLRs rely on spleen tyrosine kinase (Syk) as an adaptor molecule. Syk binds to proteins with ITAMs. Phosphorylation of the tyrosine residues in the ITAM of the receptor by kinases of the Src family creates a dock for Syk, and a further conformational change activates Syk. Activation of Syk promotes its autophosphorylation and phosphorylation of other proteins downstream in the signalling cascade [93]. Some CLRs that use Syk require ITAM-bearing adaptors that associate with them in *trans* (e.g. FcRγ or DAP12); others can bind Syk directly through a single tyrosine-based motif in the intracellular domain. This domain has been named hemITAM [93].

Dectin-1 is the prototypical example for the Syk-coupled receptors with a hemITAM. It has been postulated that upon ligand recognition by dectin 1, dimerisation of the receptors occurs, bringing together two hemITAMs that serve as a docking site for Syk; dectin localises to specific lipid microdomains which are essential for signalling [71]. In myeloid cells dectin-1 uses the adaptor CARD9 to couple Syk signalling to NFκB activation. In humans, dectin-1 signalling triggers the formation of a protein complex that includes CARD9–Bcl10–MALT-1 which couples dectin-1 to the canonical NFκB pathway by activating NFκB subunit p65 and c-Rel. Dectin-1 also triggers the noncanonical NFκB RelB pathway [71, 94, 95]. MALT-1 has been shown to act as a pivotal regulator of the c-Rel subunit; silencing of MALT-1 specifically abrogated c-Rel activation in human DCs stimulated with the dectin-1 ligand curdlan but did not affect the other NFκB subunits. The proteolytic paracaspase activity of MALT-1 was required for c-Rel activation [96]. The ability of MALT-1 to activate c-Rel was linked to the production of IL-1β and IL-12p19 in cells stimulated with curdlan or in response to *C. albicans* infection and also induction of Th17 responses [96].

Signalling through dectin-1 not only promotes polarisation of Th cells into Th17 cells but also contributes in the development of the Th1 phenotype as well as cytotoxic CD8⁺ cells [97–99].

Dectin 1–Syk-dependent activation of the NLRP3 inflammasome has also been reported in the context of fungal infection with *C. albicans*. Whereas pro-IL-1β synthesis is a

Syk–CARD9-dependent process, NLRP3 inflammasome activation requires ROS production and K⁺ efflux. In agreement with these results, mice deficient in NLRP3 are more susceptible to *C. albicans* infection supporting a role for the inflammasome in antifungal responses [80].

As mentioned before, dectin-1 can also trigger the noncanonical activation of NFκB RelB subunit. This activity requires the kinase Raf-1. While Syk activates both canonical and noncanonical pathways, Raf-1 activation triggers acetylation of the NFκB p65 subunit which can modulate transcription in association with p50. Alternatively, acetylated p65 can bind the RelB activated by Syk to render it inactive. This results in negative regulation of the RelB-dependent cytokines that include IL-23p19, hence potentiating IL-12p70 formation, which in turn favours Th1-biased responses [95]. Additionally, dectin-1 signalling triggers activation of p38, ERK and JNK pathways as well as nuclear factor of activated T cells (NFAT), an inducible nuclear factor that binds the IL-2 promoter in activated T cells [71]. Activation of NFAT by dectin-1 agonists induces secretion of a particular set of cytokines in DCs combining proinflammatory cytokines together with high levels of IL-10 and IL-2 [71, 100]. Dectin-2 also signals through the Syk pathway, but since it lacks an intracellular signalling motif, dectin-2 associates with ITAM-containing FcRγ chains [101]. Although dectin-2 also triggers NFκB activation, it does it by selectively activating c-Rel through the recruitment of MALT-1, resulting in secretion of IL-1β and IL-23, important Th17 polarising cytokines. Dectin-2 has also been shown to trigger ERK, JNK and p38MAPK pathways in murine DCs [71].

3.2.2 ITIM-Expressing CLRs

Little is known about ITIM-bearing CLRs and the signalling pathways downstream of these receptors. Several human and mouse ITIM-expressing CLRs have been reported in immune cells including DCIR (DC-inhibitory receptor), MICL (myeloid inhibitory C-type lectin receptor), CLEC12B and Ly49Q [71, 93]. It has been proposed that activation of ITIM-bearing CLRs has a regulatory effect on myeloid cells, raising the threshold for cell activation.

DCIR has been shown to inhibit TLR signalling. Specifically, it has been shown that production of IFN-I upon stimulation of TLR9 or induction of IL-12 and TNF-α through TLR8 ligation is downregulated when DCIR is cross-linked with antibodies [93]. It has been proposed that activation of DCIR is followed by phosphorylation of the ITIM domain, leading to the recruitment of SHP-I and SHP-2, two phosphatases that inhibit TLR-dependent NFκB activation [71]. Similar findings have been reported for Ly49Q and other ITIM-expressing CLRs. However, signalling pathways have not yet been fully elucidated for these receptors, and the effect of their activation still needs to be addressed in vivo.

3.2.3 CLRs Without ITAM or ITIM Domains

Some CLRs engage signalling pathways that function independently of ITIM and ITAM domains. These signalling pathways seem to play a role in regulation and fine tuning of cells activated through other receptors rather than themselves acting as a trigger for cell activation. MR, DEC-205, DC-SIGN, SIGNR and langerin are some of the receptors that engage ITAM-/ITIM-independent signalling pathways [71].

DC-SIGN has been used as a model for ITAM-/ITIM-independent CLR signalling. DC-SIGN is involved in endocytosis of soluble ligands and particulates. Besides its role in endocytosis, DC-SIGN can trigger signalling cascades and act in coordination with other PRR like TLRs. It has been shown that DC-SIGN can modulate signalling triggered by TLR ligands. Particularly, binding of the *M. tuberculosis*-derived mannosylated lipoarabinomannan (ManLAM) to DC-SIGN impairs LPS-induced maturation of DCs and increases the production of the immunosuppressive cytokine IL-10. Although the nature of the ligand appears to regulate the outcome of DC-SIGN-mediated responses, it seems that the receptor has the ability to act as an immunomodulator. It has been shown that recognition of ManLAM, by DC-SIGN, leads to activation of a signalling complex that triggers the threonine–serine kinase Raf-1 which in turn mediates acetylation of the NFκB subunit p65. When TLR signalling is triggered, acetylation of p65 mediated by DC-SIGN-Raf-1 prolongs transcriptional activity of NFκB particularly enhancing IL-10 gene transcription and thus modulating TLR responses. IL-10 induction through DC-SIGN and TLR-dependent pathways was observed for several mycobacteria such as *M. tuberculosis*, *M. leprae* and *M. bovis*, BCG and also for *C. albicans* [102].

4   RIG-I-Like Receptors

The family of RIG-I-like receptors (RLRs) consists of a small group of cytosolic receptors that act as sensors of viral RNA. So far three members have been described: retinoic acid-inducible gene-I (RIG-I), its homolog the melanoma differentiation-associated gene 5 (MDA5) and laboratory of genetics and physiology 2 (LGP2) [103, 104]. Together with endosomal TLRs, the RIG-I-like receptors detect nucleic acids inside the cells, but in contrast to TLRs, which function in the lumen of endosomes, RLRs are located in the cytosol. Hence, RLRs sense pathogens that successfully bypassed detection in the extracellular or endosomal compartments and reached the cytosol. In contrast to TLRs that are mainly expressed in immune cells, RLRs are constitutively expressed in a wide variety of immune and nonimmune cells including epithelial cells of the central nervous system. Expression of RLRs is normally maintained at low levels in resting cells but is inducible by viral infections, IFN-I stimulation and after TLR signalling in a IFN-I-independent fashion [105, 106].

4.1 Structure of RLRs

The three members of the RLR family show conserved structure and domain organisation, particularly in the case of RIG-I and MDA5. The receptors are organised in three different domains. The N-terminal region of MDA5 and RIG-I but not LGP2 contains a caspase recruitment and activation domains (CARD). Because LPG2 lacks CARD domains, it was considered an inhibitory receptor, but it was later found to play a positive role in MDA5 signalling. A central domain harbouring a DExD/H box RNA helicase that hydrolyses ATP and binds RNA is found in all three receptors. The C-terminal domain is involved in regulation and is partially responsible for ligand specificity [105, 107].

4.2 Ligands

RIG-I and MDA5 have been reported to recognise viruses from different families including Paramyxoviridae, Flaviviridae, Rhabdiviridae and Picornaviridae. RIG-I has been implicated in recognition of hepatitis C virus (HCV), Sendai and Newcastle viruses and vesicular stomatitis virus. MDA5 has been shown to be involved in recognition of poliovirus and dengue virus which is also recognised by RIG-I [107]. While LGP2 has the ability to bind RNA, its role during viral infections it is not yet clear. RIG-I was initially described as a sensor for double-stranded RNA (dsRNA), including the synthetic ligand poly(I:C) [104]. It is now clear that RIG-I recognises RNA sequences harbouring a triphosphorylated 5′ end (5′ppp). The 5′ppp end serves as a label for non-self RNA [108]. Although the length of the RNA sequence is not an absolute determinant, RIG-I has greater affinity for short RNA molecules with the 5′ppp end and dsRNA motifs [109]. 5′-hydroxyl (5′-OH) and 3′-monophosphoryl short RNA molecules with double-stranded stems generated by RNase L have been also reported to activate RIG-I, suggesting that RIG-I could recognise ligands derived from viral genomes, viral replication intermediates, viral transcripts or RNA cleaved by RNase L during infections [107]. Interestingly, RIG-I was also implicated in the sensing of dsDNA, specifically B-forms of poly(dA:dT). Sensing of poly(dA:dT) required the DNA-dependent RNA polymerase III that is able to synthesise 5′ppp RNA from poly(dA:dT) [110].

Ligands for MDA5 are less well characterised. It is known that MDA5 can be activated by poly(I:C) which suggests that it acts as a sensor for dsRNA [111]. Until now there are no reported ligands for LGP2.

4.3 Signalling Pathways Triggered by RLRs

In the absence of its ligand, RIG-I adopts an autorepressed form, preventing the CARD domains from signalling by blocking dsRNA binding to the helicase or modification of the CARD domains by ubiquitination enzymes. Binding of the RNA induces a conformational change that together with ATP hydrolysis results in release of the CARD domains, leaving them available for signalling interactions. CARDs are polyubiquitinated. This modification triggers formation of a complex formed by four RIG-I molecules [105, 107]. MDA5 has been reported to form a polar filamentous oligomer around the dsRNA ligand, which is regulated by ATP hydrolysis [112].

Signalling downstream of the complex formed by RIG-I relies on the mitochondrial protein MAVS (mitochondrial antiviral signalling), a protein that also functions as an adaptor used by the NLR member NOD2 in the context of viral infections. MAVS is anchored to the mitochondrial outer membrane and harbours an N-terminal CARD domain that allows it to establish homotypic interactions with the CARD domains of RIG-I and also MDA5 [113, 114]. During viral infections, MAVS aggregates and localises to the outer mitochondrial membrane. Together with maintenance of the mitochondrial membrane potential, these events are critical for induction of IRF3 translocation and production of IFN-I [115].

Signalling downstream of MAVS involves NEMO, IKK and TBK1. Ubiquitination is critical for downstream signalling, and it is sensed by NEMO through its ubiquitin-binding domains, allowing recruitment of IKK and TBK1 which in turn phosphorylate IκBα and IRF3, respectively, promoting translocation of IRF3 and the NFκB subunits p50 and p65 [107]. Different members of the E3 ligases, namely, TRAF6, TRAF2 and TRAF5, have been reported to be recruited to MAVS complexes and participate in the antiviral responses elicited by RIG-I engagement [107].

Besides promoting innate responses, RLRs play an important role in modulating cell-mediated immunity. IFN-I secretion promotes maturation of APCs and expression of MHC class I molecules in most cell types and is required to promote T-cell survival and expansion. It has been also proposed that interferons can promote the cytolytic activity of cytotoxic T lymphocytes and natural killer cells, enhancing the antiviral response [116, 117].

In conclusion RLRs are important PRRs in the context of viral infections. RLR activation seems to be crucial for the onset of antiviral responses and also serves to upregulate expression of other PRRs including TLRs. Enhancement of TLR expression by RLR has been shown to have an impact on the MyD88 signalling pathway and to play an important role in some viral infections. Cross talk between RLRs and other PRRs can also enhance inflammation, having negative consequences for the host [105]. Coordinated cross-regulation of the different signalling pathways is not only important for infection control but also for preventing exacerbation of inflammation and damage to the host.

5   NOD-Like Receptor Family

The members of the nucleotide-binding oligomerisation domain (NOD)-like (NLR) family have emerged as pivotal sensors of infection and stress in intracellular compartments, capable of orchestrating innate immunity and inflammation in response to harmful signals within the cell. NLRs detect a wide range of signals including the presence of intracellular PAMPs that function as flags for cellular invasion; other NLRs are activated following loss of cell membrane integrity, ion imbalance, radical oxygen species (ROS) or sensing of extracellular ATP [118].

NLRs have the ability to activate NFκB signalling; some of them function as scaffolds for the formation of a multiprotein complex known as inflammasomes required for the generation of bioactive IL-1β and IL-18 and can also trigger cell death by a mechanism known as pyroptosis [118].

Although the primary physiological role of NLRs is related to host defence against infection, in the last decade, it became increasingly clear that NLRs play a vital role in homeostasis as illustrated by many inflammatory and noninflammatory diseases that are linked to dysregulated NLR signalling [119].

Vertebrate NLRs have been the subject of intense research, while knowledge on invertebrates is limited, probably in part due to the absence of NLRs in invertebrate model organisms like *D. melanogaster* and *Caenorhabditis elegans* [120, 121]. In any case, studies revealing that NLRs are conserved across different species and kingdoms suggest they are an essential product of evolution. This is consistent with their conservation from sponges to humans and the finding that plants also express NB-LRR receptors with remarkable structural and functional similarities, although the relation of animal NLRs and the latter seems to be a result of convergent evolution rather than shared ancestry [121]. In the following section structure, the ligands and function of NLRs are discussed, with a particular focus on non-inflammasome-related NLRs, whereas inflammasome-forming NLRs are introduced in the following chapters.

5.1   Structure and Triggers of NLRs

NLRs are cytosolic sensors for microbes, endogenous danger signals and exogenous insults. The defining feature of NLR family members is the presence of a nucleotide-binding domain, the NACHT domain (acronym standing for NAIP (neuronal apoptosis inhibitor protein), CIITA (class II transcription activator), HET-E and TP-1 (telomerase-associated protein)) and a second domain harbouring a leucine-rich repeat [122]. Based on in silico studies, 22 human and 34 murine NLRs have been identified so far [123]. All the NLRs share common structural features and are organised in three functional domains including the aforementioned central NACHT (or NBD) domain necessary for oligomerisation, the C-terminal LRR that confers ligand recognition specificity and an N-terminal protein–protein interaction domain required for signal transduction [123].

The NLRs are subdivided into four subfamilies according to the type of N-terminal effector domain: NLRA, NLRB, NLRC and NLRP. The NLRA subfamily contains only one member named CIITA which presents an acidic transactivation domain [122] involved in transcriptional regulation of the MHC class II genes [123]. The other three subfamilies are characterised by the presence of homotypic protein–protein interaction modules that are involved in recruitment of signal transduction molecules. The NLRB subfamily is distinguished by the presence of the baculovirus inhibitor repeat (BIR) domain. The presence of a CARD is a feature of the NLRC subfamily, while members of the NLRP family contain a pyrin domain (PYD). Finally NLRX1, a CARD-related X effector domain of unknown function shows no strong homology to the N-terminal domain of any other NLR subfamily member [122]. Members of each subfamily are listed in Table 3.

NLRs detect a wide range of ligands of diverse origins. As in the case of TLRs, NLRs can detect PAMPs, particularly when these PAMPs reach the cytosol. Different bacterial components such as bacteria muramyl dipeptide (MDP) [124, 125], flagellin [126] and bacterial secretion systems [127, 128] function as flags for cellular invasion and trigger NLR activation and signalling. NLRs can also be activated by stress or danger signals including loss of cell membrane integrity as induced by certain bacterial toxins that form pores in the cell membrane [122, 129, 130]. Endogenous signals of damage also act as triggers of NLRs, such as membrane rupture caused by insoluble crystals [131], extracellular ATP [132–134], ion imbalance [135–137] and reactive oxygen species (ROS) [118]. Sensing the loss of cellular integrity allows NLRs to act as a backup system if a pathogen has bypassed detection in the extracellular space and also ensures recognition of DAMPs. Accordingly, the cellular inflammatory programme following NLR triggering is complex and varies between NLR members.

The NLRs have been shown to function as scaffolds for inflammasome formation. These are high molecular weight oligomeric complexes that act as caspase 1-activating platforms in response to microbial components or sterile danger and stress signals. Inflammasome complexes are formed by a sensor molecule, often but not exclusively, a member of the NLR family which connects to caspase 1 via an adaptor protein named ASC (apoptosis-associated speck-like protein containing a caspase recruitment domain) equipped with two death-fold domains (pyrin domain and caspase activation and recruitment domain (CARD)). The adaptor ASC interacts with the inflammasome sensor molecules via the pyrin

Table 3
NOD-like receptors nomenclature

NLR member	Family	Domain structure
CTIIA	NLRA	(CARD)-AD-NACHT-NADLRR
NAIP	NLRB	(CARD)-AD-NACHT-NADLRR
NOD1	NLRC	(CARD)-AD-NACHT-NADLRR
NOD2	NLRC	CARD2x-NACHT-NAD-LRR
NLRC3	NLRC	CARD-NACHT-NAD-LR
NLRC4	NLRC	CARD-NACHT-NAD-LR
NLRC5	NLRC	CARD-NACHT-NAD-LR
NLRP1	NLRP	PYD-NACHT-NAD-LRR-FIIND-CARD
NLRP2	NLRP	PYD-NACHT-NAD-LRR
NLRP3	NLRP	PYD-NACHT-NAD-LRR
NLRP4	NLRP	PYD-NACHT-NAD-LRR
NLRP5	NLRP	PYD-NACHT-NAD-LRR
NLRP6	NLRP	PYD-NACHT-NAD-LRR
NLRP7	NLRP	PYD-NACHT-NAD-LRR
NLRP8	NLRP	PYD-NACHT-NAD-LRR
NLRP9	NLRP	PYD-NACHT-NAD-LRR
NLRP10	NLRP	PYD-NACHT-NAD-LRR
NLRP11	NLRP	PYD-NACHT-NAD-LRR
NLRP12	NLRP	PYD-NACHT-NAD-LRR
NLRP13	NLRP	PYD-NACHT-NAD-LRR
NLRP14	NLRP	PYD-NACHT-NAD-LRR
NLRX1	NLX	X-NACHT-NAD-LRR

AD acidic activation domain, *CARD* caspase activating and recruitment domain, *LRR* leucine-rich repeat, *NACHT* NAIP (neuronal apoptosis inhibitor protein), C2TA (MHC class 2 transcription activator), HET-E (incompatibility locus protein from Podospora anserina) and TP1 (telomerase-associated protein), *PYD* pyrin domain, *NAD* NACHT-associated domain

domain and triggers the assembly of a large protein speck consisting mainly of multimers of ASC dimers. The CARD domains recruit caspase 1 to induce self-cleavage and activation, which in turn will allow processing of pro-IL-1β and pro-IL-18 into the active inflammatory forms and their release via a nonclassical secretion pathway [138].

The increasing number of studies on NLRs in the context of inflammasome formation in the last decade illustrates the important and varied roles of these complexes in immunology, linking inflammasome formation not only to antimicrobial responses but also autoimmunity. The following chapters will discuss in detail the role of NLRs as part of inflammasomes. However, several members of the mammalian NLR family exert important roles in immunity beyond inflammasome signalling. Here we highlight the emerging roles of several members of the non-inflammasome NLRs, CIITA, NOD1, NOD2, NLRC3, NLRC5 and NLRX1.

5.1.1 CIITA

CIITA plays a critical role in immune responses, acting as a transcriptional coactivator that regulates major histocompatibility complex (MHC) class I and II genes. Its importance as a regulator of the MHC genes was identified after finding that patients with an autoimmune condition known as bare lymphocyte syndrome have a 24 amino acid deletion splice mutant of CIITA [139]. Depending on the cell type, CIITA can be constitutively expressed (e.g. in DCs, macrophages and other cells with high MHC II expression) or can be induced by IFN-γ in a wide range of cell types [140].

In addition to the conventional tripartite architecture of NLRs, CIITA harbours three additional N-terminal domains including an acidic domain (AD), a guanosine-binding domain (GB-domain) and a Pro-Ser-Thr domain (PST domain). Although CIITA does not bind DNA directly, the AD and GBD domains mediate interactions with transcription factors, DNA-binding transactivators and chromatin-remodelling enzymes forming a complex known as the enhanceosome [123].

Although NLRs are mostly cytoplasmic receptors, CIITA can also reside in the nucleus [119]. A nuclear localisation signal is present in the GB-domain that allows trafficking into the nucleus [123].

CIITA function is regulated by phosphorylation. Several protein kinases (PK) such as PKA, PKC, glycogen synthase kinase 3 and casein kinase 2 can phosphorylate CIITA on different sites affecting its activity. CIITA possesses acetyltransferase (AT) and kinase activities, both of which are needed for effective transcription of MHC class I and II genes [123]. Although CIITA has been primarily characterized as a transcriptional regulator of MHC genes, it also regulates transcription of over 60 immunologically important genes, including IL-4, IL-10 and several thyroid-specific genes [140, 141].

Given the distinctive immunomodulatory role of CIITA, the potential of other NLR members to exert comparable effects is being addressed. It was recently proposed that NLRC5 can act as a class I transactivator. NLRC5 presents a similar domain structure to CIITA, and it has been proposed to assemble a multiprotein complex similar to the enhanceosome on MHC class I promoters [142].

5.1.2 NOD1 and NOD2

NOD1 and NOD2 were the first members of the NLR family to be described. Also known as CARD4 and CARD15, NOD1 and NOD2 were first described as receptors for LPS. However, it was later confirmed that the LPS preparations used for the experiments were contaminated by peptidoglycan moieties [119].

Gram-positive and Gram-negative bacteria synthesise peptido-glycan although they may present different motifs. Both Gram-positive and Gram-negative bacteria express muramyl dipeptide (MDP) (MurNAc-L-Ala-D-γ-Gln), but only Gram-negative bacteria and a limited number of Gram-positive bacteria express iEDAP (γ-D-glutamyl-meso-diaminopimelic acid); when it also includes the D-alanine residue from the peptidoglycan, the ligand is termed TriDAP. The difference between MDP and iEDAP is the replacement of the meso-diaminopimelic acid in the iEDAP by an L-lysine residue in MDP (Fig. 3a). NOD1 recognises the iEDAP/TriDAP, while NOD2 recognises MDP [124, 143, 144]. More recently, N-glycolyl MDP was shown to be a more potent activator of NOD2 [145]. The fact that iEDAP is mainly expressed in Gram-negative bacteria led to the idea that NOD1 was acting as a sensor for this type of microorganism. Indeed, NOD1 was shown to be involved in recognition of many different Gram-negative bacteria including *Helicobacter pylori* [146], *Pseudomonas aeruginosa* [147] and *Shigella flexneri* [148]. However, NOD1 has also been implicated in defence against Gram-positive bacteria including *Listeria monocytogenes* [149] and *Streptococcus pneumoniae* in a model of coinfection with Gram-negative bacteria [150, 151] and more surprisingly against the parasite *Trypanosoma cruzi*, etiological agent of Chagas disease [152].

NOD2 is seen as a more general sensor because its ligand MDP is widely expressed. Experimental evidence has confirmed a role for NOD2 during infection with *M. tuberculosis* [145], *Listeria monocytogenes* [153] and *Toxoplasma gondii* [154]. NOD2 may also play a role in antiviral immunity; NOD2-deficient mice exhibited a marked susceptibility to respiratory syncytial virus infection compared to the wild-type counterparts [155].

NOD2 has been also linked to pathology in Crohn's disease, an inflammatory disease that mainly affects the ileum and colon. An increased susceptibility to developing this condition was linked to several mutations in NOD2, although the aetiology is not fully understood. It has also been postulated that NOD2 could negatively regulates TLR-mediated inflammation since NOD2 deficiency or a mutation related to Crohn's disease increased Toll-like receptor 2-mediated activation of NFκB and Th1 responses. Moreover, NOD2 inhibited TLR2-driven activation of NFκB [156].

NOD1 and NOD2 expression has been reported in a wide variety of cells including dendritic cells [157], monocytes/macrophages [158], keratinocytes [159], lung and intestinal epithelial cells [160, 161] and endothelial cells [162]. Although several cell types constitutively express NOD1 and NOD2, its expression can also be induced in response to cytokines [163], TLR ligands and bacteria [164, 165]. Signalling through NOD1 and NOD2 ultimately triggers NFκB activation and MAPK. The IRF pathway and IFN-I transcription can be also triggered by NOD receptors (Fig. 3b).

The first step in the signalling cascade involves dimerisation of the receptors [119]. Although both are considered cytosolic receptors, association with the plasma membrane seems to occur after ligand binding [166].

It has been proposed that bacterial peptidoglycans are internalised in endosomes and can access the cytosol by exiting the vesicles through channels including hPepT1 and SLC15A. Scavenger receptors such as MARCO and SR-A have been implicated in rapid internalisation of NOD ligands [123]. After recognition of the ligand and assembly of the dimers, a protein adaptor known as RICK/RIP2 is an ubiquitinated interaction with the CARD domains of NOD1 or NOD2 [167]. RICK participates in the recruitment of TAK1; it also promotes ubiquitination of IκKγ which acts as a regulator of the IκK complex. NEMO, another regulator of NFκB, is also recruited and facilitates TAK1 recruitment to the complex. Formation of this complex promotes phosphorylation and targeting of the IκK complex subunits for proteasomal degradation, promoting release of NFκB [119, 123]. Activation of NOD1 and NOD2 also triggers activation of the MAPK pathway although this has received less attention.

Finally, NOD receptors have been shown to activate the IRF pathway (Fig. 3b). NOD2 induces type-I IFN secretion upon recognition of viral single-stranded RNA or during infections with respiratory syncytial virus and influenza virus [155]. Induction of IFN-I involves activation of a RICK-independent pathway with formation of a complex with the protein MAVS, the adaptor used by RLRs during viral infections [118].

NOD1 has been shown to induce IFN-I production after recognition of iE-DAP. Binding of iE-DAP to NOD1 triggered RICK signalling and also the recruitment of TRAF3 which was shown to trigger TBK1, IKKε and activation of IRF7, inducing IFN-β production. This ultimately led to activation of the transcription factor complex ISGF3 and secretion of CXCL-10 and further production of IFN-I [168].

As seen for TLRs, activation of the NFκB pathway through NOD1 and NOD2 receptors promotes expression of several

Fig. 3 (continued) of the IκK complex. NEMO, another regulator of NFκB, is also recruited and facilitates TAK1 recruitment to the complex. Formation of this complex promotes phosphorylation, and targeting of the IκK complex subunits for proteasomal degradation promoting release of NFκB of NOD1 and NOD2 also triggers activation of the MAPK pathway. The type-I IFN pathway can also be activated by NOD receptors. NOD2 induces type-I IFN secretion after recognising viral single-stranded RNA or during viral infections with respiratory syncytial virus and influenza virus. A RICK-independent pathway is triggered for IFN-I production. This pathway involves formation of a complex with the protein MAVS. Translocation of IRF3 and IRF7 takes place

a

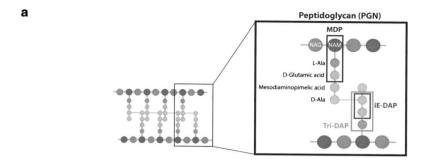

b

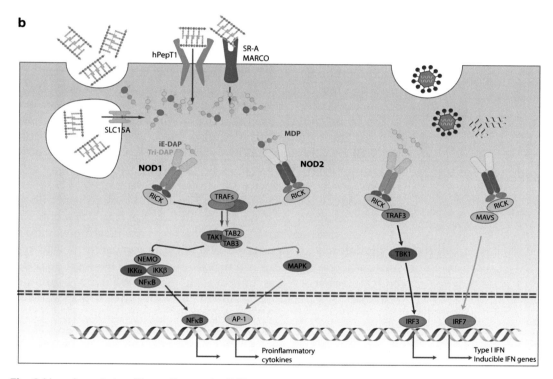

Fig. 3 Ligands and signalling pathways for NOD1 and NOD2. (**a**) *Structure of peptidoglycan (PGN) of Gram-negative bacteria*. PGN is composed of the alternating amino sugars N-acetylglucosamine (NAG) and N-acetyl muramic acid (NAM) cross-linked by ß1-4 linkages. NAG and NAM units are cross-linked by stem peptides containing amino acids such as D-glutamic acid and D- or L-alanine. Generally, the third position amino acid in Gram-positive bacteria is L-lysine (Lys), while in Gram-negative bacteria it is meso-2,6-diaminopimelic (meso-DAP) acid. Gram-positive bacteria have peptide stems usually cross-linked through an interpeptide bridge (normally glycine), whereas Gram-negative bacteria peptide stems are usually directly cross-linked. Abbreviations: *iE-DAP* D-g-glutamyl-meso-DAP, *Tri-Dap* L-Ala-γ-D-Glu-mDAP, *MDP* muramyl dipeptide. (**b**) *Signalling pathways for NOD1 and NOD2*. Transport of bacterial PNG fragments is mediated by endosomal SLC15A channel, the hPepT1 plasma membrane transporter or scavenger receptors (SR-A, MARCO). MDP and iE-DAP from PNG activates NOD1 and NOD2, respectively. Direct or indirect sensing of PNG leads to NOD1 and NOD2 dimerisation and relocalisation to the plasma membrane and triggers and recruitment of the adaptor RICK. RICK participates in recruitment of TAK1; it also promotes ubiquitination of IκKγ which acts as a regulator

inflammatory factors including inducible nitric oxide synthase, cyclooxygenase 2, adhesion molecules and proinflammatory cytokines and chemokines such as such as TNF-α, IL-8, IL-6 [118, 119]. Relatively little is known about how NOD1 and NOD2 signalling pathways are regulated. It has been suggested that the ubiquitin-editing enzyme A20 plays a key role in regulating RICK activity upon MDP recognition by NOD2. A20 deficiency amplified the responses to MDP in cells by increasing RICK ubiquitination, resulting in prolonged NFκB signalling and increased production of proinflammatory cytokines. A similar phenotype has been found in A20-deficient mice [169]. Caspase 12 has been also implicated in negative regulation of the NOD signalling pathway by binding to RIPK2 and destabilising the complex between RICK and TRAF6, ultimately inhibiting the ubiquitin-ligase activity of the complex [170].

NOD agonists have also been implicated in enhancing antigen-specific antibodies and T-cell responses when combined with TLR ligands. One example of the importance of NOD receptor in enhancing adaptive responses is given by the diminished responses observed in NOD1-deficient mice immunised with a model protein antigen (ovalbumin) formulated in complete Freund's adjuvant (CFA). CFA contains mycobacterial cell wall elements known to activate both NOD and TLR receptors. NOD1-deficient mice showed lower frequencies of antigen-specific IFN-γ, IL-17 and IL-4-producing CD4+ and CD8+ T cells, and diminished antibody titres compared to wild-type mice. Besides, increased susceptibility of NOD1-deficient mice to *Helicobacter pylori* infection was linked to diminished urease-specific IgG2c titres compared to wild-type mice, whereas IgG1 titres remained the same. This suggested that NOD1 deficiency was linked to a diminished type 1 immune response [171]. This is in agreement with other studies showing that NOD1 and NOD2 agonists in combination with TLR3, TLR4 and TLR9 agonists synergistically induce IL-12 and IFN-γ production in DCs to induce Th1-biased immune responses [157]. Another example of the potential synergistic effects of TLR-NOD on adaptive responses was given by a study showing that a chimeric NOD2/TLR2 agonist not only induced dendritic cell maturation and proinflammatory cytokine secretion in vitro but also boosted systemic and mucosal immune responses after parenteral immunisation of mice when formulated as a nanoparticle-based vaccine carrying the HIV antigen Gag p24 [172].

5.1.3 NLRC3

NLRC3 is another member of the NLR family, but instead of promoting inflammation, NLRC3 acts as a negative regulator of several innate receptors. It was first described as a negative regulator of T-cell proliferation, an effect mediated by downregulation of the transcription factors NFκB, AP-1 and NFAT. NLRC3 expression was shown to be high in T cells and downregulated upon activation [173].

The same receptor was also shown to have regulatory effects on TLR-mediated signalling by directly interacting with TRAF6 as evidenced by the increased proinflammatory cytokine production in NLRC3-deficient mice upon stimulation with LPS [174].

Lastly, NLRC3 has been shown to act as a negative regulator of the cytosolic DNA sensor STING by directly associating with it and TBK1, preventing the interaction between the two of them reducing production of IFN-I in response to cyclic diguanylate monophosphate (c-di-GMP) and DNA viruses [175]. NLRC3 illustrates how PRRs can regulate each other to fine-tune the immune response in the host.

5.1.4 NLRC5

NLRC5 is structurally similar to CIITA and is also involved in regulation of MHC genes, more specifically as a class I transactivator. It has been described as the largest member of the NLR family, harbouring an atypically long LRR motif and death domain [142]. Several isoforms of NLRC5 with unknown function have also been described [176]. Expression of NLRC5 has been reported in several cell types, mainly of haematopoietic origin. It is highly expressed in lymphoid tissues like the spleen and lymph nodes and has also been reported to be expressed in the bone marrow. NLRC5 has been documented in T cells, B cells and mononuclear cells [176].

NLRC5 promotes the expression of conventional MHC class I genes (HLA-A, HLA-B, HLA-C) as well as nonconventional MHC class I genes and proteins like HLA-E. NLRC5 regulation was shown to be exclusive for MHC class I genes and not MHC class II, whereas as previously mentioned, CTIIA is a MHC-II regulator [177].

Besides its function as a class I transactivator, NLRC5 has been implicated in negative regulation of NFκB and IRF pathways. NLRC5 can inhibit the IKK complex and RIG-I/MDA5 function. By blocking phosphorylation of IKKα and IKKβ, NLRC5 blocked the activation of NFκB. Additionally, interaction with RIG-I and MDA5 was shown to inhibit IFN-I responses triggered by RIG-like receptors [178]. Despite this, macrophages and dendritic cells derived from NLRC5-deficient mice did not exhibit compromised production of IFN-β, IL-6 or TNF-α when stimulated with RNA viruses, DNA viruses or bacteria [179]. Hence, the regulation could be cell-type specific.

5.1.5 NLRX1

This member of the NLR family has been one of the most controversial regarding its function. Although NLRX1 was first described as a negative regulator of antiviral responses, particularly by down-regulating IFN-I secretion in response to viral infections [180], other studies could not confirm this effect. Additionally, knockout mice showed normal IFN-β production after influenza A infection or systemically administered poly (I:C) [181, 182]. In other set of studies NLRX1 was shown to negatively regulate TLR-mediated activation of NFκB and JNK signalling pathways [183].

6   Newly Described PRRs

The recent discovery of NLRs and RLRs opened new avenues for studying other cytosolic receptors, leading to discovery of new PRRs. In the past decade, several proteins have been described as intracellular sensors for nucleic acids, in particular the family of AIM2-like receptors or ALRs together with the DNA sensor cGAS.

6.1   AIM2-Like Receptors

The family of AIM2-like receptors comprises four members in humans and six in mice. The founding member of this new family of PRRs is the absent in melanoma 2 (AIM2) protein that was described in 2009 [184–186]. The members of the ALR family are characterised by the presence of a HIN200 domain, also known as IFI200. Three members of the family in humans (MNDA, PYHIN1 and AIM2) present an N-terminal PYD domain, whereas the fourth member IFI16 harbours two tandem HIN200 domains and one PYD. The HIN200 domains can interact with cytosolic dsDNA, either of viral or bacterial origin, triggering IFN-I production [187]. Also, the PYD domain can recruit the adaptor ASC to form an inflammasome and lead to production of bioactive IL-1β. Importantly, until now only AIM2 and IFI16 have been reported as PRRs. The fact that ALRs have only been described in mammals suggests they are a novel family of receptors that appeared later in evolution [187].

6.2   cGAS and STING

cGAS has recently emerged as a major sensor for cytosolic DNA and together with the adaptor protein STING which is widely expressed in various cell types, they contribute to DNA sensing from different origins including DNA from viruses, self DNA and sensing of bacterial cyclic dinucleotides such as c-di-GMP and c-di-AMP [107, 188, 189]. Activation of STING triggers trafficking of the adaptor from the endoplasmic reticulum to the Golgi apparatus for the assembly of protein complexes with TBK1 [190]. The STING pathway can trigger IFN-I production and has been also reported to recruit STAT6, promoting secretion of chemokines including CCL-2, CCL-20 and CCL-26 [191].

The cyclic GMP-AMP synthase (cGAS), a member of the nucleotidyltransferase family, has been identified as a cytosolic DNA sensor that contributes to the production of IFN-I. cGAS catalyses the synthesis of cyclic GMP-AMP (cGAMP) from ATP and GTP, which in turn acts as a second messenger for STING [192, 193].

Deficiency of cGAS in vivo abolished type-I IFN production in response to cytosolic DNA [194]. cGAS signalling has proven important in the context of several viral infections caused by DNA viruses and interestingly also contributes to antiviral immunity to RNA viruses [195, 196].

7   Conclusion

Overall, the discovery of PRRs has changed the way innate immunity is viewed and has brought it back to the foreground. Unravelling the molecular mechanisms of immune recognition revealed the interconnection between innate and adaptive immunity, supporting the notion that innate recognition is a key event that allows the host to mount the most effective immune response.

Even though recognition of microbes or abnormal-self is restricted to a finite number of molecular patterns, these confer a significant degree of specificity allowing for tailored responses. During infection several receptors are simultaneously activated, triggering specific combinations of signalling pathways that converge on NFκB, MAPK and IRFs. These pathways that share common players downstream from the PRRs engage in cross-talk that can result in synergy, enhancement, negative regulation and fine-tuning of gene expression. Moreover, given that the different PRRs display differential expression patterns that can be tissue or cell-type specific, those interactions will be dependent on the context in which the activation takes place. PRRs have been shown to cross-regulate each other, and defining how this cross-talk is regulated is crucial to better understand how pathways that seem to converge on the same master regulators can facilitate different outcomes. This will certainly open new avenues for treatment of inflammatory diseases and immune deficiencies and will make it possible to exploit PRRs as a means to induce protective immunity through vaccination.

References

1. Cooper EL (2010) Evolution of immune systems from self/not self to danger to artificial immune systems (AIS). Phys Life Rev 7(1):55–78. doi:10.1016/j.plrev.2009.12.001
2. Metschnikoff E (1884) Ueber eine Sprosspilzkrankheit der Daphnien. Beitrag zur Lehre über den Kampf der Phagocyten gegen Krankheitserreger. Archiv f Pathol Anat 96(2):177–195. doi:10.1007/BF02361555
3. Wilson JDKE (2007) Sir Frank Macfarlane Burnet 1899–1985. Nat Immunol 8(10):1009. doi:10.1038/ni1007-1009
4. Landsteiner K (1933) Die Spezifität des serologischen Reaktionen. Springer, Berlin
5. Burnet FM (1959) The clonal selection theory of acquired immunity. Vanderbilt University Press, Nashville
6. Janeway CAJ (1989) Approaching the asymptote? Evolution and revolution in immunology. Cold Spring Harb Symp Quant Biol 54:1–13

7. Liu Y, Janeway CAJ (1991) Microbial induction of co-stimulatory activity for CD4 T-cell growth. Int Immunol 3(4):323–332
8. Liu Y, Janeway CAJ (1992) Cells that present both specific ligand and costimulatory activity are the most efficient inducers of clonal expansion of normal CD4 T cells. Proc Natl Acad Sci U S A 89(9):3845–3849
9. Matzinger P (1994) Tolerance, danger, and the extended family. Annu Rev Immunol 12:991–1045. doi:10.1146/annurev.iy.12.040194.005015
10. Lemaitre B, Nicolas E, Michaut L, Reichhart JM, Hoffmann JA (1996) The dorsoventral regulatory gene cassette spatzle/Toll/cactus controls the potent antifungal response in Drosophila adults. Cell 86(6):973–983
11. Medzhitov R, Preston-Hurlburt P, Janeway CA Jr (1997) A human homologue of the Drosophila Toll protein signals activation of

adaptive immunity. Nature 388(6640):394–397. doi:10.1038/41131

12. Schaefer L (2014) Complexity of danger: the diverse nature of damage-associated molecular patterns. J Biol Chem 289(51):35237–35245. doi:10.1074/jbc.R114.619304

13. Broz P, Monack DM (2013) Newly described pattern recognition receptors team up against intracellular pathogens. Nat Rev Immunol 13(8):551–565. doi:10.1038/nri3479

14. Iwasaki A, Medzhitov R (2010) Regulation of adaptive immunity by the innate immune system. Science 327(5963):291–295. doi:10.1126/science.1183021

15. Hashimoto C, Hudson KL, Anderson KV (1988) The Toll gene of Drosophila, required for dorsal-ventral embryonic polarity, appears to encode a transmembrane protein. Cell 52(2):269–279

16. Gay NJ, Keith FJ (1991) Drosophila Toll and IL-1 receptor. Nature 351(6325):355–356. doi:10.1038/351355b0

17. Dinarello CA (1991) Interleukin-1 and interleukin-1 antagonism. Blood 77(8):1627–1652

18. Heguy A, Baldari CT, Macchia G, Telford JL, Melli M (1992) Amino acids conserved in interleukin-1 receptors (IL-1Rs) and the Drosophila Toll protein are essential for IL-1R signal transduction. J Biol Chem 267(4):2605–2609

19. Sen R, Baltimore D (1986) Inducibility of kappa immunoglobulin enhancer-binding protein Nf-kappa B by a posttranslational mechanism. Cell 47(6):921–928

20. Kawai T, Akira S (2009) The roles of TLRs, RLRs and NLRs in pathogen recognition. Int Immunol 21(4):317–337. doi:10.1093/intimm/dxp017

21. Buchmann K (2014) Evolution of innate immunity: clues from invertebrates via fish to mammals. Front Immunol 5:459. doi:10.3389/fimmu.2014.00459

22. Botos I, Segal DM, Davies DR (2011) The structural biology of Toll-like receptors. Structure 19(4):447–459. doi:10.1016/j.str.2011.02.004

23. Lee BL, Moon JE, Shu JH, Yuan L, Newman ZR, Schekman R, Barton GM (2013) UNC93B1 mediates differential trafficking of endosomal TLRs. Elife 2, e00291. doi:10.7554/eLife.00291

24. Pifer R, Benson A, Sturge CR, Yarovinsky F (2011) UNC93B1 is essential for TLR11 activation and IL-12-dependent host resistance to Toxoplasma gondii. J Biol Chem 286(5):3307–3314. doi:10.1074/jbc.M110.171025

25. Jin MS, Lee JO (2008) Structures of the Toll-like receptor family and its ligand complexes. Immunity 29(2):182–191. doi:10.1016/j.immuni.2008.07.007

26. Omueti KO, Beyer JM, Johnson CM, Lyle EA, Tapping RI (2005) Domain exchange between human Toll-like receptors 1 and 6 reveals a region required for lipopeptide discrimination. J Biol Chem 280(44):36616–36625. doi:10.1074/jbc.M504320200

27. Kang JY, Nan X, Jin MS, Youn SJ, Ryu YH, Mah S, Han SH, Lee H, Paik SG, Lee JO (2009) Recognition of lipopeptide patterns by Toll-like receptor 2-Toll-like receptor 6 heterodimer. Immunity 31(6):873–884. doi:10.1016/j.immuni.2009.09.018

28. Liu L, Botos I, Wang Y, Leonard JN, Shiloach J, Segal DM, Davies DR (2008) Structural basis of Toll-like receptor 3 signaling with double-stranded RNA. Science 320(5874):379–381. doi:10.1126/science.1155406

29. Park BS, Song DH, Kim HM, Choi BS, Lee H, Lee JO (2009) The structural basis of lipopolysaccharide recognition by the TLR4-MD-2 complex. Nature 458(7242):1191–1195. doi:10.1038/nature07830

30. Akira S, Uematsu S, Takeuchi O (2006) Pathogen recognition and innate immunity. Cell 124(4):783–801. doi:10.1016/j.cell.2006.02.015

31. Matsushima N, Tanaka T, Enkhbayar P, Mikami T, Taga M, Yamada K, Kuroki Y (2007) Comparative sequence analysis of leucine-rich repeats (LRRs) within vertebrate Toll-like receptors. BMC Genomics 8:124. doi:10.1186/1471-2164-8-124

32. Yarovinsky F, Zhang D, Andersen JF, Bannenberg GL, Serhan CN, Hayden MS, Hieny S, Sutterwala FS, Flavell RA, Ghosh S, Sher A (2005) TLR11 activation of dendritic cells by a protozoan profilin-like protein. Science 308(5728):1626–1629. doi:10.1126/science.1109893

33. Koblansky AA, Jankovic D, Oh H, Hieny S, Sungnak W, Mathur R, Hayden MS, Akira S, Sher A, Ghosh S (2013) Recognition of profilin by Toll-like receptor 12 is critical for host resistance to Toxoplasma gondii. Immunity 38(1):119–130. doi:10.1016/j.immuni.2012.09.016

34. Mathur R, Oh H, Zhang D, Park SG, Seo J, Koblansky A, Hayden MS, Ghosh S (2012) A mouse model of Salmonella typhi infection. Cell 151(3):590–602. doi:10.1016/j.cell.2012.08.042

35. Zhang D, Zhang G, Hayden MS, Greenblatt MB, Bussey C, Flavell RA, Ghosh S (2004) A Toll-like receptor that prevents infection by uro-

pathogenic bacteria. Science 303(5663):1522–1526. doi:10.1126/science.1094351

36. Oldenburg M, Kruger A, Ferstl R, Kaufmann A, Nees G, Sigmund A, Bathke B, Lauterbach H, Suter M, Dreher S, Koedel U, Akira S, Kawai T, Buer J, Wagner H, Bauer S, Hochrein H, Kirschning CJ (2012) TLR13 recognizes bacterial 23S rRNA devoid of erythromycin resistance-forming modification. Science 337(6098):1111–1115. doi:10.1126/science.1220363

37. Godfroy JI, Roostan M, Moroz YS, Korendovych IV, Yin H (2012) Isolated Toll-like receptor transmembrane domains are capable of oligomerization. PLoS One 7(11), e48875. doi:10.1371/journal.pone.0048875

38. Bowie A, O'Neill LA (2000) The interleukin-1 receptor/Toll-like receptor superfamily: signal generators for pro-inflammatory interleukins and microbial products. J Leukoc Biol 67(4):508–514

39. O'Neill LA, Golenbock D, Bowie AG (2013) The history of Toll-like receptors—redefining innate immunity. Nat Rev Immunol 13(6):453–460. doi:10.1038/nri3446

40. Kawai T, Akira S (2007) Signaling to NF-kappaB by Toll-like receptors. Trends Mol Med 13(11):460–469. doi:10.1016/j.molmed.2007.09.002

41. Savitsky D, Tamura T, Yanai H, Taniguchi T (2010) Regulation of immunity and oncogenesis by the IRF transcription factor family. Cancer Immunol Immunother 59(4):489–510. doi:10.1007/s00262-009-0804-6

42. Lin SC, Lo YC, Wu H (2010) Helical assembly in the MyD88-IRAK4-IRAK2 complex in TLR/IL-1R signalling. Nature 465(7300):885–890. doi:10.1038/nature09121

43. Kawasaki T, Kawai T (2014) Toll-like receptor signaling pathways. Front Immunol 5:461. doi:10.3389/fimmu.2014.00461

44. Israel A (2010) The IKK complex, a central regulator of NF-kappaB activation. Cold Spring Harb Perspect Biol 2(3):a000158. doi:10.1101/cshperspect.a000158

45. Wang C, Deng L, Hong M, Akkaraju GR, Inoue J, Chen ZJ (2001) TAK1 is a ubiquitin-dependent kinase of MKK and IKK. Nature 412(6844):346–351. doi:10.1038/35085597

46. Eliopoulos AG, Dumitru CD, Wang CC, Cho J, Tsichlis PN (2002) Induction of COX-2 by LPS in macrophages is regulated by Tpl2-dependent CREB activation signals. EMBO J 21(18):4831–4840

47. Banerjee A, Gugasyan R, McMahon M, Gerondakis S (2006) Diverse Toll-like receptors utilize Tpl2 to activate extracellular signal-regulated kinase (ERK) in hemopoietic cells. Proc Natl Acad Sci U S A 103(9):3274–3279. doi:10.1073/pnas.0511113103

48. Takeuchi O, Kaufmann A, Grote K, Kawai T, Hoshino K, Morr M, Muhlradt PF, Akira S (2000) Cutting edge: preferentially the R-stereoisomer of the mycoplasmal lipopeptide macrophage-activating lipopeptide-2 activates immune cells through a Toll-like receptor 2- and MyD88-dependent signaling pathway. J Immunol 164(2):554–557

49. Schnare M, Holt AC, Takeda K, Akira S, Medzhitov R (2000) Recognition of CpG DNA is mediated by signaling pathways dependent on the adaptor protein MyD88. Curr Biol 10(18):1139–1142

50. Hayashi F, Smith KD, Ozinsky A, Hawn TR, Yi EC, Goodlett DR, Eng JK, Akira S, Underhill DM, Aderem A (2001) The innate immune response to bacterial flagellin is mediated by Toll-like receptor 5. Nature 410(6832):1099–1103. doi:10.1038/35074106

51. Kaisho T, Takeuchi O, Kawai T, Hoshino K, Akira S (2001) Endotoxin-induced maturation of MyD88-deficient dendritic cells. J Immunol 166(9):5688–5694

52. Adachi O, Kawai T, Takeda K, Matsumoto M, Tsutsui H, Sakagami M, Nakanishi K, Akira S (1998) Targeted disruption of the MyD88 gene results in loss of IL-1- and IL-18-mediated function. Immunity 9(1):143–150

53. Sims JE, Smith DE (2010) The IL-1 family: regulators of immunity. Nat Rev Immunol 10(2):89–102. doi:10.1038/nri2691

54. Alexopoulou L, Holt AC, Medzhitov R, Flavell RA (2001) Recognition of double-stranded RNA and activation of NF-kappaB by Toll-like receptor 3. Nature 413(6857):732–738. doi:10.1038/35099560

55. Kawai T, Adachi O, Ogawa T, Takeda K, Akira S (1999) Unresponsiveness of MyD88-deficient mice to endotoxin. Immunity 11(1):115–122

56. Kawai T, Takeuchi O, Fujita T, Inoue J, Muhlradt PF, Sato S, Hoshino K, Akira S (2001) Lipopolysaccharide stimulates the MyD88-independent pathway and results in activation of IFN-regulatory factor 3 and the expression of a subset of lipopolysaccharide-inducible genes. J Immunol 167(10): 5887–5894

57. Oshiumi H, Matsumoto M, Funami K, Akazawa T, Seya T (2003) TICAM-1, an adaptor molecule that participates in Toll-like receptor 3-mediated interferon-beta induc-

tion. Nat Immunol 4(2):161–167. doi:10.1038/ni886

58. Yamamoto M, Sato S, Hemmi H, Hoshino K, Kaisho T, Sanjo H, Takeuchi O, Sugiyama M, Okabe M, Takeda K, Akira S (2003) Role of adaptor TRIF in the MyD88-independent Toll-like receptor signaling pathway. Science 301(5633):640–643. doi:10.1126/science.1087262

59. Yamamoto M, Sato S, Mori K, Hoshino K, Takeuchi O, Takeda K, Akira S (2002) Cutting edge: a novel Toll/IL-1 receptor domain-containing adapter that preferentially activates the IFN-beta promoter in the Toll-like receptor signaling. J Immunol 169(12):6668–6672

60. Kawai T, Akira S (2007) TLR signaling. Semin Immunol 19(1):24–32. doi:10.1016/j.smim.2006.12.004

61. Carty M, Goodbody R, Schroder M, Stack J, Moynagh PN, Bowie AG (2006) The human adaptor SARM negatively regulates adaptor protein TRIF-dependent Toll-like receptor signaling. Nat Immunol 7(10):1074–1081. doi:10.1038/ni1382

62. Honda K, Yanai H, Negishi H, Asagiri M, Sato M, Mizutani T, Shimada N, Ohba Y, Takaoka A, Yoshida N, Taniguchi T (2005) IRF-7 is the master regulator of type-I interferon-dependent immune responses. Nature 434(7034):772–777. doi:10.1038/nature03464

63. Cao W, Manicassamy S, Tang H, Kasturi SP, Pirani A, Murthy N, Pulendran B (2008) Toll-like receptor-mediated induction of type I interferon in plasmacytoid dendritic cells requires the rapamycin-sensitive PI(3)K-mTOR-p70S6K pathway. Nat Immunol 9(10):1157–1164. doi:10.1038/ni.1645

64. Kawai T, Sato S, Ishii KJ, Coban C, Hemmi H, Yamamoto M, Terai K, Matsuda M, Inoue J, Uematsu S, Takeuchi O, Akira S (2004) Interferon-alpha induction through Toll-like receptors involves a direct interaction of IRF7 with MyD88 and TRAF6. Nat Immunol 5(10):1061–1068. doi:10.1038/ni1118

65. Shinohara ML, Lu L, Bu J, Werneck MB, Kobayashi KS, Glimcher LH, Cantor H (2006) Osteopontin expression is essential for interferon-alpha production by plasmacytoid dendritic cells. Nat Immunol 7(5):498–506. doi:10.1038/ni1327

66. Honda K, Ohba Y, Yanai H, Negishi H, Mizutani T, Takaoka A, Taya C, Taniguchi T (2005) Spatiotemporal regulation of MyD88-IRF-7 signalling for robust type-I interferon induction. Nature 434(7036):1035–1040. doi:10.1038/nature03547

67. Tsujimura H, Tamura T, Kong HJ, Nishiyama A, Ishii KJ, Klinman DM, Ozato K (2004) Toll-like receptor 9 signaling activates NF-kappaB through IFN regulatory factor-8/IFN consensus sequence binding protein in dendritic cells. J Immunol 172(11):6820–6827

68. Tailor P, Tamura T, Kong HJ, Kubota T, Kubota M, Borghi P, Gabriele L, Ozato K (2007) The feedback phase of type I interferon induction in dendritic cells requires interferon regulatory factor 8. Immunity 27(2):228–239. doi:10.1016/j.immuni.2007.06.009

69. Takaoka A, Yanai H, Kondo S, Duncan G, Negishi H, Mizutani T, Kano S, Honda K, Ohba Y, Mak TW, Taniguchi T (2005) Integral role of IRF-5 in the gene induction programme activated by Toll-like receptors. Nature 434(7030):243–249. doi:10.1038/nature03308

70. Negishi H, Fujita Y, Yanai H, Sakaguchi S, Ouyang X, Shinohara M, Takayanagi H, Ohba Y, Taniguchi T, Honda K (2006) Evidence for licensing of IFN-gamma-induced IFN regulatory factor 1 transcription factor by MyD88 in Toll-like receptor-dependent gene induction program. Proc Natl Acad Sci U S A 103(41):15136–15141. doi:10.1073/pnas.0607181103

71. Sancho D, Reis e Sousa C (2012) Signaling by myeloid C-type lectin receptors in immunity and homeostasis. Annu Rev Immunol 30:491–529. doi:10.1146/annurev-immunol-031210-101352

72. Drummond RA, Brown GD (2013) Signalling C-type lectins in antimicrobial immunity. PLoS Pathog 9(7), e1003417. doi:10.1371/journal.ppat.1003417

73. Zelensky AN, Gready JE (2005) The C-type lectin-like domain superfamily. FEBS J 272(24):6179–6217. doi:10.1111/j.1742-4658.2005.05031.x

74. Hoving JC, Wilson GJ, Brown GD (2014) Signalling C-type lectin receptors, microbial recognition and immunity. Cell Microbiol 16(2):185–194. doi:10.1111/cmi.12249

75. Iborra S, Sancho D (2015) Signalling versatility following self and non-self sensing by myeloid C-type lectin receptors. Immunobiology 220(2):175–184. doi:10.1016/j.imbio.2014.09.013

76. Geijtenbeek TB, Gringhuis SI (2009) Signalling through C-type lectin receptors: shaping immune responses. Nat Rev Immunol 9(7):465–479. doi:10.1038/nri2569

77. Dambuza IM, Brown GD (2015) C-type lectins in immunity: recent developments. Curr Opin Immunol 32C:21–27. doi:10.1016/j.coi.2014.12.002

78. Gringhuis SI, Kaptein TM, Wevers BA, Theelen B, van der Vlist M, Boekhout T, Geijtenbeek TB (2012) Dectin-1 is an extracellular pathogen sensor for the induction and processing of IL-1beta via a noncanonical caspase-8 inflammasome. Nat Immunol 13(3):246–254. doi:10.1038/ni.2222

79. Hise AG, Tomalka J, Ganesan S, Patel K, Hall BA, Brown GD, Fitzgerald KA (2009) An essential role for the NLRP3 inflammasome in host defense against the human fungal pathogen *Candida albicans*. Cell Host Microbe 5(5):487–497. doi:10.1016/j.chom.2009.05.002

80. Gross O, Poeck H, Bscheider M, Dostert C, Hannesschlager N, Endres S, Hartmann G, Tardivel A, Schweighoffer E, Tybulewicz V, Mocsai A, Tschopp J, Ruland J (2009) Syk kinase signalling couples to the Nlrp3 inflammasome for anti-fungal host defence. Nature 459(7245):433–436. doi:10.1038/nature07965

81. del Fresno C, Soulat D, Roth S, Blazek K, Udalova I, Sancho D, Ruland J, Ardavin C (2013) Interferon-beta production via Dectin-1-Syk-IRF5 signaling in dendritic cells is crucial for immunity to *C. albicans*. Immunity 38(6):1176–1186. doi:10.1016/j.immuni.2013.05.010

82. Dorhoi A, Desel C, Yeremeev V, Pradl L, Brinkmann V, Mollenkopf HJ, Hanke K, Gross O, Ruland J, Kaufmann SH (2010) The adaptor molecule CARD9 is essential for tuberculosis control. J Exp Med 207(4):777–792. doi:10.1084/jem.20090067

83. Zamze S, Martinez-Pomares L, Jones H, Taylor PR, Stillion RJ, Gordon S, Wong SY (2002) Recognition of bacterial capsular polysaccharides and lipopolysaccharides by the macrophage mannose receptor. J Biol Chem 277(44):41613–41623. doi:10.1074/jbc.M207057200

84. Astarie-Dequeker C, N'Diaye EN, Le Cabec V, Rittig MG, Prandi J, Maridonneau-Parini I (1999) The mannose receptor mediates uptake of pathogenic and nonpathogenic mycobacteria and bypasses bactericidal responses in human macrophages. Infect Immun 67(2):469–477

85. Schulert GS, Allen LA (2006) Differential infection of mononuclear phagocytes by *Francisella tularensis*: role of the macrophage mannose receptor. J Leukoc Biol 80(3):563–571. doi:10.1189/jlb.0306219

86. Zhang SS, Park CG, Zhang P, Bartra SS, Plano GV, Klena JD, Skurnik M, Hinnebusch BJ, Chen T (2008) Plasminogen activator Pla of Yersinia pestis utilizes murine DEC-205

(CD205) as a receptor to promote dissemination. J Biol Chem 283(46):31511–31521. doi:10.1074/jbc.M804646200

87. Geijtenbeek TB, van Kooyk Y (2003) DC-SIGN: a novel HIV receptor on DCs that mediates HIV-1 transmission. Curr Top Microbiol Immunol 276:31–54

88. Hillaire ML, Nieuwkoop NJ, Boon AC, de Mutsert G, Vogelzang-van Trierum SE, Fouchier RA, Osterhaus AD, Rimmelzwaan GF (2013) Binding of DC-SIGN to the hemagglutinin of influenza A viruses supports virus replication in DC-SIGN expressing cells. PLoS One 8(2), e56164. doi:10.1371/journal.pone.0056164

89. Iborra S, Izquierdo HM, Martínez-López M, Blanco-Menéndez N, Reis e Sousa C, Sancho D (2012) The DC receptor DNGR-1 mediates cross-priming of CTLs during vaccinia virus infection in mice. J Clin Invest 122(5):1628–1643. doi:10.1172/JCI60660

90. Zelenay S, Keller AM, Whitney PG, Schraml BU, Deddouche S, Rogers NC, Schulz O, Sancho D, Reis e Sousa C (2012) The dendritic cell receptor DNGR-1 controls endocytic handling of necrotic cell antigens to favor cross-priming of CTLs in virus-infected mice. J Clin Invest 122(5):1615–1627. doi:10.1172/JCI60644

91. van Die I, van Vliet SJ, Nyame AK, Cummings RD, Bank CM, Appelmelk B, Geijtenbeek TB, van Kooyk Y (2003) The dendritic cell-specific C-type lectin DC-SIGN is a receptor for Schistosoma mansoni egg antigens and recognizes the glycan antigen Lewis x. Glycobiology 13(6):471–478. doi:10.1093/glycob/cwg052

92. Ritter M, Gross O, Kays S, Ruland J, Nimmerjahn F, Saijo S, Tschopp J, Layland LE, Prazeres da Costa C (2010) *Schistosoma mansoni* triggers Dectin-2, which activates the Nlrp3 inflammasome and alters adaptive immune responses. Proc Natl Acad Sci U S A 107(47):20459–20464. doi:10.1073/pnas.1010337107

93. Osorio F, Reis e Sousa C (2011) Myeloid C-type lectin receptors in pathogen recognition and host defense. Immunity 34(5):651–664. doi:10.1016/j.immuni.2011.05.001

94. Goodridge HS, Shimada T, Wolf AJ, Hsu YM, Becker CA, Lin X, Underhill DM (2009) Differential use of CARD9 by dectin-1 in macrophages and dendritic cells. J Immunol 182(2):1146–1154

95. Gringhuis SI, den Dunnen J, Litjens M, van der Vlist M, Wevers B, Bruijns SC, Geijtenbeek TB (2009) Dectin-1 directs T helper cell differentiation by controlling noncanonical

NF-kappaB activation through Raf-1 and Syk. Nat Immunol 10(2):203–213. doi:10.1038/ni.1692

96. Gringhuis SI, Wevers BA, Kaptein TM, van Capel TM, Theelen B, Boekhout T, de Jong EC, Geijtenbeek TB (2011) Selective C-Rel activation via Malt1 controls anti-fungal T(H)-17 immunity by dectin-1 and dectin-2. PLoS Pathog 7(1), e1001259. doi:10.1371/journal.ppat.1001259

97. Leibundgut-Landmann S, Gross O, Robinson MJ, Osorio F, Slack EC, Tsoni SV, Schweighoffer E, Tybulewicz V, Brown GD, Ruland J, Reis e Sousa C (2007) Syk- and CARD9-dependent coupling of innate immunity to the induction of T helper cells that produce interleukin 17. Nat Immunol 8(6):630–638. doi:10.1038/ni1460

98. Leibundgut-Landmann S, Osorio F, Brown GD, Reis e Sousa C (2008) Stimulation of dendritic cells via the dectin-1/Syk pathway allows priming of cytotoxic T-cell responses. Blood 112(13):4971–4980. doi:10.1182/blood-2008-05-158469

99. Osorio F, LeibundGut-Landmann S, Lochner M, Lahl K, Sparwasser T, Eberl G, Reis e Sousa C (2008) DC activated via dectin-1 convert Treg into IL-17 producers. Eur J Immunol 38(12):3274–3281. doi:10.1002/eji.200838950

100. Goodridge HS, Simmons RM, Underhill DM (2007) Dectin-1 stimulation by *Candida albicans* yeast or zymosan triggers NFAT activation in macrophages and dendritic cells. J Immunol 178(5):3107–3115

101. Sato K, Yang XL, Yudate T, Chung JS, Wu J, Luby-Phelps K, Kimberly RP, Underhill D, Cruz PD Jr, Ariizumi K (2006) Dectin-2 is a pattern recognition receptor for fungi that couples with the Fc receptor gamma chain to induce innate immune responses. J Biol Chem 281(50):38854–38866. doi:10.1074/jbc.M606542200

102. Gringhuis SI, den Dunnen J, Litjens M, van Het Hof B, van Kooyk Y, Geijtenbeek TB (2007) C-type lectin DC-SIGN modulates Toll-like receptor signaling via Raf-1 kinase-dependent acetylation of transcription factor NF-kappaB. Immunity 26(5):605–616. doi:10.1016/j.immuni.2007.03.012

103. Yoneyama M, Kikuchi M, Matsumoto K, Imaizumi T, Miyagishi M, Taira K, Foy E, Loo YM, Gale M Jr, Akira S, Yonehara S, Kato A, Fujita T (2005) Shared and unique functions of the DExD/H-box helicases RIG-I, MDA5, and LGP2 in antiviral innate immunity. J Immunol 175(5):2851–2858

104. Yoneyama M, Kikuchi M, Natsukawa T, Shinobu N, Imaizumi T, Miyagishi M, Taira K, Akira S, Fujita T (2004) The RNA helicase RIG-I has an essential function in double-stranded RNA-induced innate antiviral responses. Nat Immunol 5(7):730–737. doi:10.1038/ni1087

105. Loo YM, Gale M Jr (2011) Immune signaling by RIG-I-like receptors. Immunity 34(5):680–692. doi:10.1016/j.immuni.2011.05.003

106. Szabo A, Magyarics Z, Pazmandi K, Gopcsa L, Rajnavolgyi E, Bacsi A (2014) TLR ligands upregulate RIG-I expression in human plasmacytoid dendritic cells in a type I IFN-independent manner. Immunol Cell Biol 92(8):671–678. doi:10.1038/icb.2014.38

107. Wu J, Chen ZJ (2014) Innate immune sensing and signaling of cytosolic nucleic acids. Annu Rev Immunol 32:461–488. doi:10.1146/annurev-immunol-032713-120156

108. Hornung V, Ellegast J, Kim S, Brzozka K, Jung A, Kato H, Poeck H, Akira S, Conzelmann KK, Schlee M, Endres S, Hartmann G (2006) 5'-Triphosphate RNA is the ligand for RIG-I. Science 314(5801):994–997. doi:10.1126/science.1132505

109. Baum A, Sachidanandam R, Garcia-Sastre A (2010) Preference of RIG-I for short viral RNA molecules in infected cells revealed by next-generation sequencing. Proc Natl Acad Sci U S A 107(37):16303–16308. doi:10.1073/pnas.1005077107

110. Chiu YH, Macmillan JB, Chen ZJ (2009) RNA polymerase III detects cytosolic DNA and induces type I interferons through the RIG-I pathway. Cell 138(3):576–591. doi:10.1016/j.cell.2009.06.015

111. Kato H, Takeuchi O, Sato S, Yoneyama M, Yamamoto M, Matsui K, Uematsu S, Jung A, Kawai T, Ishii KJ, Yamaguchi O, Otsu K, Tsujimura T, Koh CS, Reis e Sousa C, Matsuura Y, Fujita T, Akira S (2006) Differential roles of MDA5 and RIG-I helicases in the recognition of RNA viruses. Nature 441(7089):101–105. doi:10.1038/nature04734

112. Peisley A, Lin C, Wu B, Orme-Johnson M, Liu M, Walz T, Hur S (2011) Cooperative assembly and dynamic disassembly of MDA5 filaments for viral dsRNA recognition. Proc Natl Acad Sci U S A 108(52):21010–21015. doi:10.1073/pnas.1113651108

113. Seth RB, Sun L, Ea CK, Chen ZJ (2005) Identification and characterization of MAVS, a mitochondrial antiviral signaling protein that activates NF-kappaB and IRF 3.

Cell 122(5):669–682. doi:10.1016/j.cell.2005.08.012

114. Sun Q, Sun L, Liu HH, Chen X, Seth RB, Forman J, Chen ZJ (2006) The specific and essential role of MAVS in antiviral innate immune responses. Immunity 24(5):633–642. doi:10.1016/j.immuni.2006.04.004

115. Hou F, Sun L, Zheng H, Skaug B, Jiang QX, Chen ZJ (2011) MAVS forms functional prion-like aggregates to activate and propagate antiviral innate immune response. Cell 146(3):448–461. doi:10.1016/j.cell.2011.06.041

116. Biron CA, Nguyen KB, Pien GC, Cousens LP, Salazar-Mather TP (1999) Natural killer cells in antiviral defense: function and regulation by innate cytokines. Annu Rev Immunol 17:189–220. doi:10.1146/annurev.immunol.17.1.189

117. Curtsinger JM, Valenzuela JO, Agarwal P, Lins D, Mescher MF (2005) Type I IFNs provide a third signal to CD8 T cells to stimulate clonal expansion and differentiation. J Immunol 174(8):4465–4469

118. Kersse K, Bertrand MJ, Lamkanfi M, Vandenabeele P (2011) NOD-like receptors and the innate immune system: coping with danger, damage and death. Cytokine Growth Factor Rev 22(5–6):257–276. doi:10.1016/j.cytogfr.2011.09.003

119. Chen G, Shaw MH, Kim YG, Nunez G (2009) NOD-like receptors: role in innate immunity and inflammatory disease. Annu Rev Pathol 4:365–398. doi:10.1146/annurev.pathol.4.110807.092239

120. Bryant CE, Monie TP (2012) Mice, men and the relatives: cross-species studies underpin innate immunity. Open Biol 2(4):120015. doi:10.1098/rsob.120015

121. Yuen B, Bayes JM, Degnan SM (2014) The characterization of sponge NLRs provides insight into the origin and evolution of this innate immune gene family in animals. Mol Biol Evol 31(1):106–120. doi:10.1093/molbev/mst174

122. Ting JP, Lovering RC, Alnemri ES, Bertin J, Boss JM, Davis BK, Flavell RA, Girardin SE, Godzik A, Harton JA, Hoffman HM, Hugot JP, Inohara N, Mackenzie A, Maltais LJ, Nunez G, Ogura Y, Otten LA, Philpott D, Reed JC, Reith W, Schreiber S, Steimle V, Ward PA (2008) The NLR gene family: a standard nomenclature. Immunity 28(3):285–287. doi:10.1016/j.immuni.2008.02.005

123. Barbe F, Douglas T, Saleh M (2014) Advances in Nod-like receptors (NLR) biology. Cytokine Growth Factor Rev 25(6):681–697. doi:10.1016/j.cytogfr.2014.07.001

124. Girardin SE, Boneca IG, Carneiro LA, Antignac A, Jehanno M, Viala J, Tedin K, Taha MK, Labigne A, Zahringer U, Coyle AJ, DiStefano PS, Bertin J, Sansonetti PJ, Philpott DJ (2003) Nod1 detects a unique muropeptide from gram-negative bacterial peptidoglycan. Science 300(5625):1584–1587. doi:10.1126/science.1084677

125. Tanabe T, Chamaillard M, Ogura Y, Zhu L, Qiu S, Masumoto J, Ghosh P, Moran A, Predergast MM, Tromp G, Williams CJ, Inohara N, Nunez G (2004) Regulatory regions and critical residues of NOD2 involved in muramyl dipeptide recognition. EMBO J 23(7):1587–1597. doi:10.1038/sj.emboj.7600175

126. Zhao Y, Yang J, Shi J, Gong YN, Lu Q, Xu H, Liu L, Shao F (2011) The NLRC4 inflammasome receptors for bacterial flagellin and type III secretion apparatus. Nature 477(7366):596–600. doi:10.1038/nature10510

127. Rayamajhi M, Zak DE, Chavarria-Smith J, Vance RE, Miao EA (2013) Cutting edge: mouse NAIP1 detects the type III secretion system needle protein. J Immunol 191(8):3986–3989. doi:10.4049/jimmunol.1301549

128. Yang J, Zhao Y, Shi J, Shao F (2013) Human NAIP and mouse NAIP1 recognize bacterial type III secretion needle protein for inflammasome activation. Proc Natl Acad Sci U S A 110(35):14408–14413. doi:10.1073/pnas.1306376110

129. Mariathasan S, Weiss DS, Newton K, McBride J, O'Rourke K, Roose-Girma M, Lee WP, Weinrauch Y, Monack DM, Dixit VM (2006) Cryopyrin activates the inflammasome in response to toxins and ATP. Nature 440(7081):228–232. doi:10.1038/nature04515

130. McNeela EA, Burke A, Neill DR, Baxter C, Fernandes VE, Ferreira D, Smeaton S, El-Rachkidy R, McLoughlin RM, Mori A, Moran B, Fitzgerald KA, Tschopp J, Petrilli V, Andrew PW, Kadioglu A, Lavelle EC (2010) Pneumolysin activates the NLRP3 inflammasome and promotes proinflammatory cytokines independently of TLR4. PLoS Pathog 6(11), e1001191. doi:10.1371/journal.ppat.1001191

131. Martinon F, Petrilli V, Mayor A, Tardivel A, Tschopp J (2006) Gout-associated uric acid crystals activate the NALP3 inflammasome. Nature 440(7081):237–241. doi:10.1038/nature04516

132. Duncan JA, Bergstralh DT, Wang Y, Willingham SB, Ye Z, Zimmermann AG, Ting JP (2007) Cryopyrin/NALP3 binds ATP/dATP, is an ATPase, and requires ATP

binding to mediate inflammatory signaling. Proc Natl Acad Sci U S A 104(19):8041–8046. doi:10.1073/pnas.0611496104

133. Ye Z, Lich JD, Moore CB, Duncan JA, Williams KL, Ting JP (2008) ATP binding by monarch-1/NLRP12 is critical for its inhibitory function. Mol Cell Biol 28(5):1841–1850. doi:10.1128/MCB.01468-07

134. Zurek B, Proell M, Wagner RN, Schwarzenbacher R, Kufer TA (2012) Mutational analysis of human NOD1 and NOD2 NACHT domains reveals different modes of activation. Innate Immun 18(1):100–111. doi:10.1177/1753425910394002

135. Perregaux D, Gabel CA (1994) Interleukin-1 beta maturation and release in response to ATP and nigericin. Evidence that potassium depletion mediated by these agents is a necessary and common feature of their activity. J Biol Chem 269(21):15195–15203

136. Schorn C, Frey B, Lauber K, Janko C, Strysio M, Keppeler H, Gaipl US, Voll RE, Springer E, Munoz LE, Schett G, Herrmann M (2011) Sodium overload and water influx activate the NALP3 inflammasome. J Biol Chem 286(1):35–41. doi:10.1074/jbc.M110.139048

137. Arlehamn CS, Petrilli V, Gross O, Tschopp J, Evans TJ (2010) The role of potassium in inflammasome activation by bacteria. J Biol Chem 285(14):10508–10518. doi:10.1074/jbc.M109.067298

138. Latz E, Xiao TS, Stutz A (2013) Activation and regulation of the inflammasomes. Nat Rev Immunol 13(6):397–411. doi:10.1038/nri3452

139. Steimle V, Otten LA, Zufferey M, Mach B (1993) Complementation cloning of an MHC class II transactivator mutated in hereditary MHC class II deficiency (or bare lymphocyte syndrome). Cell 75(1):135–146

140. Nagarajan UM, Bushey A, Boss JM (2002) Modulation of gene expression by the MHC class II transactivator. J Immunol 169(9):5078–5088

141. Mori-Aoki A, Pietrarelli M, Nakazato M, Caturegli P, Kohn LD, Suzuki K (2000) Class II transactivator suppresses transcription of thyroid-specific genes. Biochem Biophys Res Commun 278(1):58–62. doi:10.1006/bbrc.2000.3769

142. Meissner TB, Li A, Kobayashi KS (2012) NLRC5: a newly discovered MHC class I transactivator (CITA). Microbes Infect 14(6):477–484. doi:10.1016/j.micinf.2011.12.007

143. Chamaillard M, Hashimoto M, Horie Y, Masumoto J, Qiu S, Saab L, Ogura Y, Kawasaki A, Fukase K, Kusumoto S, Valvano MA, Foster SJ, Mak TW, Nunez G, Inohara N (2003) An essential role for NOD1 in host recognition of bacterial peptidoglycan containing diaminopimelic acid. Nat Immunol 4(7):702–707. doi:10.1038/ni945

144. Girardin SE, Boneca IG, Viala J, Chamaillard M, Labigne A, Thomas G, Philpott DJ, Sansonetti PJ (2003) Nod2 is a general sensor of peptidoglycan through muramyl dipeptide (MDP) detection. J Biol Chem 278(11):8869–8872. doi:10.1074/jbc.C200651200

145. Coulombe F, Divangahi M, Veyrier F, de Leseleuc L, Gleason JL, Yang Y, Kelliher MA, Pandey AK, Sassetti CM, Reed MB, Behr MA (2009) Increased NOD2-mediated recognition of N-glycolyl muramyl dipeptide. J Exp Med 206(8):1709–1716. doi:10.1084/jem.20081779

146. Viala J, Chaput C, Boneca IG, Cardona A, Girardin SE, Moran AP, Athman R, Memet S, Huerre MR, Coyle AJ, DiStefano PS, Sansonetti PJ, Labigne A, Bertin J, Philpott DJ, Ferrero RL (2004) Nod1 responds to peptidoglycan delivered by the Helicobacter pylori cag pathogenicity island. Nat Immunol 5(11):1166–1174. doi:10.1038/ni1131

147. Travassos LH, Carneiro LA, Girardin SE, Boneca IG, Lemos R, Bozza MT, Domingues RC, Coyle AJ, Bertin J, Philpott DJ, Plotkowski MC (2005) Nod1 participates in the innate immune response to *Pseudomonas aeruginosa*. J Biol Chem 280(44):36714–36718. doi:10.1074/jbc.M501649200

148. Girardin SE, Tournebize R, Mavris M, Page AL, Li X, Stark GR, Bertin J, DiStefano PS, Yaniv M, Sansonetti PJ, Philpott DJ (2001) CARD4/Nod1 mediates NF-kappaB and JNK activation by invasive Shigella flexneri. EMBO Rep 2(8):736–742. doi:10.1093/embo-reports/kve155

149. Opitz B, Puschel A, Beermann W, Hocke AC, Forster S, Schmeck B, van Laak V, Chakraborty T, Suttorp N, Hippenstiel S (2006) *Listeria monocytogenes* activated p38 MAPK and induced IL-8 secretion in a nucleotide-binding oligomerization domain 1-dependent manner in endothelial cells. J Immunol 176(1):484–490

150. Lysenko ES, Clarke TB, Shchepetov M, Ratner AJ, Roper DI, Dowson CG, Weiser JN (2007) Nod1 signaling overcomes resistance of *S. pneumoniae* to opsonophagocytic killing. PLoS Pathog 3(8), e118. doi:10.1371/journal.ppat.0030118

151. Ratner AJ, Aguilar JL, Shchepetov M, Lysenko ES, Weiser JN (2007) Nod1 mediates cytoplasmic sensing of combinations of extracellular

bacteria. Cell Microbiol 9(5):1343–1351. doi:10.1111/j.1462-5822.2006.00878.x

152. Silva GK, Gutierrez FR, Guedes PM, Horta CV, Cunha LD, Mineo TW, Santiago-Silva J, Kobayashi KS, Flavell RA, Silva JS, Zamboni DS (2010) Cutting edge: nucleotide-binding oligomerization domain 1-dependent responses account for murine resistance against *Trypanosoma cruzi* infection. J Immunol 184(3):1148–1152. doi:10.4049/jimmunol.0902254

153. Kobayashi KS, Chamaillard M, Ogura Y, Henegariu O, Inohara N, Nunez G, Flavell RA (2005) Nod2-dependent regulation of innate and adaptive immunity in the intestinal tract. Science 307(5710):731–734. doi:10.1126/science.1104911

154. Shaw MH, Reimer T, Sanchez-Valdepenas C, Warner N, Kim YG, Fresno M, Nunez G (2009) T cell-intrinsic role of Nod2 in promoting type 1 immunity to Toxoplasma gondii. Nat Immunol 10(12):1267–1274. doi:10.1038/ni.1816

155. Sabbah A, Chang TH, Harnack R, Frohlich V, Tominaga K, Dube PH, Xiang Y, Bose S (2009) Activation of innate immune antiviral responses by Nod2. Nat Immunol 10(10):1073–1080. doi:10.1038/ni.1782

156. Watanabe T, Kitani A, Murray PJ, Strober W (2004) NOD2 is a negative regulator of Toll-like receptor 2-mediated T helper type 1 responses. Nat Immunol 5(8):800–808. doi:10.1038/ni1092

157. Tada H, Aiba S, Shibata K, Ohteki T, Takada H (2005) Synergistic effect of Nod1 and Nod2 agonists with Toll-like receptor agonists on human dendritic cells to generate interleukin-12 and T helper type 1 cells. Infect Immun 73(12):7967–7976. doi:10.1128/IAI.73.12.7967-7976.2005

158. Ogura Y, Inohara N, Benito A, Chen FF, Yamaoka S, Nunez G (2001) Nod2, a Nod1/Apaf-1 family member that is restricted to monocytes and activates NF-kappaB. J Biol Chem 276(7):4812–4818. doi:10.1074/jbc.M008072200

159. Voss E, Wehkamp J, Wehkamp K, Stange EF, Schroder JM, Harder J (2006) NOD2/CARD15 mediates induction of the antimicrobial peptide human beta-defensin-2. J Biol Chem 281(4):2005–2011. doi:10.1074/jbc.M511044200

160. Hisamatsu T, Suzuki M, Reinecker HC, Nadeau WJ, McCormick BA, Podolsky DK (2003) CARD15/NOD2 functions as an antibacterial factor in human intestinal epithelial cells. Gastroenterology 124(4):993–1000. doi:10.1053/gast.2003.50153

161. Uehara A, Fujimoto Y, Fukase K, Takada H (2007) Various human epithelial cells express functional Toll-like receptors, NOD1 and NOD2 to produce anti-microbial peptides, but not proinflammatory cytokines. Mol Immunol 44(12):3100–3111. doi:10.1016/j.molimm.2007.02.007

162. Opitz B, Forster S, Hocke AC, Maass M, Schmeck B, Hippenstiel S, Suttorp N, Krull M (2005) Nod1-mediated endothelial cell activation by *Chlamydophila pneumoniae*. Circ Res 96(3):319–326. doi:10.1161/01.RES.0000155721.83594.2c

163. Rosenstiel P, Fantini M, Brautigam K, Kuhbacher T, Waetzig GH, Seegert D, Schreiber S (2003) TNF-alpha and IFN-gamma regulate the expression of the NOD2 (CARD15) gene in human intestinal epithelial cells. Gastroenterology 124(4):1001–1009. doi:10.1053/gast.2003.50157

164. Kim YG, Park JH, Shaw MH, Franchi L, Inohara N, Nunez G (2008) The cytosolic sensors Nod1 and Nod2 are critical for bacterial recognition and host defense after exposure to Toll-like receptor ligands. Immunity 28(2):246–257. doi:10.1016/j.immuni.2007.12.012

165. Pudla M, Kananurak A, Limposuwan K, Sirisinha S, Utaisincharoen P (2011) Nucleotide-binding oligomerization domain-containing protein 2 regulates suppressor of cytokine signaling 3 expression in *Burkholderia pseudomallei*-infected mouse macrophage cell line RAW 264.7. Innate Immun 17(6):532–540. doi:10.1177/1753425910385484

166. Barnich N, Aguirre JE, Reinecker HC, Xavier R, Podolsky DK (2005) Membrane recruitment of NOD2 in intestinal epithelial cells is essential for nuclear factor-kappa B activation in muramyl dipeptide recognition. J Cell Biol 170(1):21–26. doi:10.1083/jcb.200502153

167. Hasegawa M, Fujimoto Y, Lucas PC, Nakano H, Fukase K, Nunez G, Inohara N (2008) A critical role of RICK/RIP2 polyubiquitination in Nod-induced NF-kappaB activation. EMBO J 27(2):373–383. doi:10.1038/sj.emboj.7601962

168. Watanabe T, Asano N, Fichtner-Feigl S, Gorelick PL, Tsuji Y, Matsumoto Y, Chiba T, Fuss IJ, Kitani A, Strober W (2010) NOD1 contributes to mouse host defense against *Helicobacter pylori* via induction of type I IFN and activation of the ISGF3 signaling pathway. J Clin Invest 120(5):1645–1662. doi:10.1172/JCI39481

169. Hitotsumatsu O, Ahmad RC, Tavares R, Wang M, Philpott D, Turer EE, Lee BL, Shiffin N, Advincula R, Malynn BA, Werts C,

Ma A (2008) The ubiquitin-editing enzyme A20 restricts nucleotide-binding oligomerization domain containing 2-triggered signals. Immunity 28(3):381–390. doi:10.1016/j.immuni.2008.02.002

170. LeBlanc PM, Yeretssian G, Rutherford N, Doiron K, Nadiri A, Zhu L, Green DR, Gruenheid S, Saleh M (2008) Caspase-12 modulates NOD signaling and regulates antimicrobial peptide production and mucosal immunity. Cell Host Microbe 3(3):146–157. doi:10.1016/j.chom.2008.02.004

171. Fritz JH, Le Bourhis L, Sellge G, Magalhaes JG, Fsihi H, Kufer TA, Collins C, Viala J, Ferrero RL, Girardin SE, Philpott DJ (2007) Nod1-mediated innate immune recognition of peptidoglycan contributes to the onset of adaptive immunity. Immunity 26(4):445–459. doi:10.1016/j.immuni.2007.03.009

172. Pavot V, Rochereau N, Resseguier J, Gutjahr A, Genin C, Tiraby G, Perouzel E, Lioux T, Vernejoul F, Verrier B, Paul S (2014) Cutting edge: new chimeric NOD2/TLR2 adjuvant drastically increases vaccine immunogenicity. J Immunol 193(12):5781–5785. doi:10.4049/jimmunol.1402184

173. Conti BJ, Davis BK, Zhang J, O'Connor W Jr, Williams KL, Ting JP (2005) CATERPILLER 16.2 (CLR16.2), a novel NBD/LRR family member that negatively regulates T cell function. J Biol Chem 280(18):18375–18385. doi:10.1074/jbc.M413169200

174. Schneider M, Zimmermann AG, Roberts RA, Zhang L, Swanson KV, Wen H, Davis BK, Allen IC, Holl EK, Ye Z, Rahman AH, Conti BJ, Eitas TK, Koller BH, Ting JP (2012) The innate immune sensor NLRC3 attenuates Toll-like receptor signaling via modification of the signaling adaptor TRAF6 and transcription factor NF-kappaB. Nat Immunol 13(9):823–831. doi:10.1038/ni.2378

175. Zhang L, Mo J, Swanson KV, Wen H, Petrucelli A, Gregory SM, Zhang Z, Schneider M, Jiang Y, Fitzgerald KA, Ouyang S, Liu ZJ, Damania B, Shu HB, Duncan JA, Ting JP (2014) NLRC3, a member of the NLR family of proteins, is a negative regulator of innate immune signaling induced by the DNA sensor STING. Immunity 40(3):329–341. doi:10.1016/j.immuni.2014.01.010

176. Neerincx A, Lautz K, Menning M, Kremmer E, Zigrino P, Hosel M, Buning H, Schwarzenbacher R, Kufer TA (2010) A role for the human nucleotide-binding domain, leucine-rich repeat-containing family member NLRC5 in antiviral responses. J Biol Chem 285(34):26223–26232. doi:10.1074/jbc.M110.109736

177. Meissner TB, Li A, Biswas A, Lee KH, Liu YJ, Bayir E, Iliopoulos D, van den Elsen PJ, Kobayashi KS (2010) NLR family member NLRC5 is a transcriptional regulator of MHC class I genes. Proc Natl Acad Sci U S A 107(31):13794–13799. doi:10.1073/pnas.1008684107

178. Cui J, Zhu L, Xia X, Wang HY, Legras X, Hong J, Ji J, Shen P, Zheng S, Chen ZJ, Wang RF (2010) NLRC5 negatively regulates the NF-kappaB and type I interferon signaling pathways. Cell 141(3):483–496. doi:10.1016/j.cell.2010.03.040

179. Kumar H, Pandey S, Zou J, Kumagai Y, Takahashi K, Akira S, Kawai T (2011) NLRC5 deficiency does not influence cytokine induction by virus and bacteria infections. J Immunol 186(2):994–1000. doi:10.4049/jimmunol.1002094

180. Allen IC, Moore CB, Schneider M, Lei Y, Davis BK, Scull MA, Gris D, Roney KE, Zimmermann AG, Bowzard JB, Ranjan P, Monroe KM, Pickles RJ, Sambhara S, Ting JP (2011) NLRX1 protein attenuates inflammatory responses to infection by interfering with the RIG-I-MAVS and TRAF6-NF-kappaB signaling pathways. Immunity 34(6):854–865. doi:10.1016/j.immuni.2011.03.026

181. Rebsamen M, Vazquez J, Tardivel A, Guarda G, Curran J, Tschopp J (2011) NLRX1/NOD5 deficiency does not affect MAVS signalling. Cell Death Differ 18(8):1387. doi:10.1038/cdd.2011.64

182. Soares F, Tattoli I, Wortzman ME, Arnoult D, Philpott DJ, Girardin SE (2013) NLRX1 does not inhibit MAVS-dependent antiviral signalling. Innate Immun 19(4):438–448. doi:10.1177/1753425912467383

183. Xia X, Cui J, Wang HY, Zhu L, Matsueda S, Wang Q, Yang X, Hong J, Songyang Z, Chen ZJ, Wang RF (2011) NLRX1 negatively regulates TLR-induced NF-kappaB signaling by targeting TRAF6 and IKK. Immunity 34(6):843–853. doi:10.1016/j.immuni.2011.02.022

184. Fernandes-Alnemri T, Yu JW, Datta P, Wu J, Alnemri ES (2009) AIM2 activates the inflammasome and cell death in response to cytoplasmic DNA. Nature 458(7237):509–513. doi:10.1038/nature07710

185. Hornung V, Ablasser A, Charrel-Dennis M, Bauernfeind F, Horvath G, Caffrey DR, Latz E, Fitzgerald KA (2009) AIM2 recognizes cytosolic dsDNA and forms a caspase-1-activating inflammasome with ASC. Nature

458(7237):514–518. doi:10.1038/nature07725

186. Roberts TL, Idris A, Dunn JA, Kelly GM, Burnton CM, Hodgson S, Hardy LL, Garceau V, Sweet MJ, Ross IL, Hume DA, Stacey KJ (2009) HIN-200 proteins regulate caspase activation in response to foreign cytoplasmic DNA. Science 323(5917):1057–1060. doi:10.1126/science.1169841

187. Hansen JD, Vojtech LN, Laing KJ (2011) Sensing disease and danger: a survey of vertebrate PRRs and their origins. Dev Comp Immunol 35(9):886–897. doi:10.1016/j.dci.2011.01.008

188. Ishikawa H, Barber GN (2008) STING is an endoplasmic reticulum adaptor that facilitates innate immune signalling. Nature 455(7213):674–678. doi:10.1038/nature07317

189. Ouyang S, Song X, Wang Y, Ru H, Shaw N, Jiang Y, Niu F, Zhu Y, Qiu W, Parvatiyar K, Li Y, Zhang R, Cheng G, Liu ZJ (2012) Structural analysis of the STING adaptor protein reveals a hydrophobic dimer interface and mode of cyclic di-GMP binding. Immunity 36(6):1073–1086. doi:10.1016/j.immuni.2012.03.019

190. Tanaka Y, Chen ZJ (2012) STING specifies IRF3 phosphorylation by TBK1 in the cytosolic DNA signaling pathway. Sci Signal 5(214), ra20. doi:10.1126/scisignal.2002521

191. Chen H, Sun H, You F, Sun W, Zhou X, Chen L, Yang J, Wang Y, Tang H, Guan Y, Xia W, Gu J, Ishikawa H, Gutman D, Barber G, Qin Z, Jiang Z (2011) Activation of STAT6 by STING is critical for antiviral innate immunity. Cell 147(2):436–446. doi:10.1016/j.cell.2011.09.022

192. Sun L, Wu J, Du F, Chen X, Chen ZJ (2013) Cyclic GMP-AMP synthase is a cytosolic DNA sensor that activates the type I interferon pathway. Science 339(6121):786–791. doi:10.1126/science.1232458

193. Zhang X, Shi H, Wu J, Zhang X, Sun L, Chen C, Chen ZJ (2013) Cyclic GMP-AMP containing mixed phosphodiester linkages is an endogenous high-affinity ligand for STING. Mol Cell 51(2):226–235. doi:10.1016/j.molcel.2013.05.022

194. Gao D, Wu J, Wu YT, Du F, Aroh C, Yan N, Sun L, Chen ZJ (2013) Cyclic GMP-AMP synthase is an innate immune sensor of HIV and other retroviruses. Science 341(6148):903–906. doi:10.1126/science.1240933

195. Li XD, Wu J, Gao D, Wang H, Sun L, Chen ZJ (2013) Pivotal roles of cGAS-cGAMP signaling in antiviral defense and immune adjuvant effects. Science 341(6152):1390–1394. doi:10.1126/science.1244040

196. Schoggins JW, MacDuff DA, Imanaka N, Gainey MD, Shrestha B, Eitson JL, Mar KB, Richardson RB, Ratushny AV, Litvak V, Dabelic R, Manicassamy B, Aitchison JD, Aderem A, Elliott RM, Garcia-Sastre A, Racaniello V, Snijder EJ, Yokoyama WM, Diamond MS, Virgin HW, Rice CM (2014) Pan-viral specificity of IFN-induced genes reveals new roles for cGAS in innate immunity. Nature 505(7485):691–695. doi:10.1038/nature12862

Chapter 2

Atypical Inflammasomes

Ann M. Janowski and Fayyaz S. Sutterwala

Abstract

Pattern recognition receptors, including members of the NLR and PYHIN families, are essential for recognition of both pathogen- and host-derived danger signals. A number of molecules in these families are capable of forming multiprotein complexes termed inflammasomes that result in the activation of caspase-1. In addition to NLRP1, NLRP3, NLRC4, and AIM2, which form well-described inflammasome complexes, IFI16, NLRP6, NLRP7, NLRP12, and NLRC5 have also been proposed to form inflammasomes under specific conditions. The structure and function of these atypical inflammasomes will be highlighted here.

Key words NLR, Caspase-1, Inflammasome

1 Introduction

Much of the inflammasome literature is focused around four distinct inflammasomes. These include the NLRP1, NLRP3, NLRC4, and AIM2 inflammasomes [1]. The structure, function, and ligand recognition of these inflammasomes is an area of intense investigation, with NLRP3 being one of the most studied due to its contribution in the immune response to a wide variety of pathogens and diseases. However, the NLR and PYHIN family of proteins are quite large, and there are a handful of additional molecules that form multiprotein complexes that differ from the four typical inflammasomes. Included in the family of atypical inflammasomes are IFI16, NLRP6, NLRP7, NLRP12, and NLRC5. For the purposes of this chapter, an inflammasome is defined as a multiprotein complex that functions as a platform for caspase-1 activation. These atypical inflammasomes have a vast array of functions and also demonstrate that inflammasome biology is still a budding area of research. This review will shine a spotlight on atypical inflammasomes and their contribution to the host immune response.

Francesco Di Virgilio and Pablo Pelegrín (eds.), *NLR Proteins: Methods and Protocols*, Methods in Molecular Biology, vol. 1417, DOI 10.1007/978-1-4939-3566-6_2, © Springer Science+Business Media New York 2016

2   IFI16: A Nuclear Inflammasome

Pattern recognition receptors are capable of recognizing and binding conserved sequences associated with pathogens (pathogen-associated molecular patterns, PAMPs) and also endogenous danger signals (danger-associated molecular patterns, DAMPs). Members of the IFI20X/IFI16 (PYHIN) family of proteins play important roles in recognizing the presence of cytosolic DNA. There are currently six identified PYHIN proteins in mice (p202a, p202b, p203, p204, MNDAL, and AIM2) and four in humans (IFI16, AIM2, MNDA, and IFIX) [2]. PYHIN family members are characterized by the presence of an N-terminal pyrin domain, which allows for homotypic interactions with other pyrin-containing proteins to form multimolecular complexes. PYHIN family members also possess a DNA-binding HIN-200 domain [2]. There are three different subtypes of HIN domains: HIN A, HIN B, and HIN C. The PYHIN family member AIM2 (absent in melanoma 2) is capable of recruiting the adaptor molecule ASC and caspase-1 to form an inflammasome. Through the sensing of cytosolic dsDNA, the AIM2 inflammasome is crucial for protection against pathogens such as *Listeria monocytogenes* and *Francisella tularensis* [3–6].

AIM2 is not the only PYHIN family member that associates with ASC and mediates protection against pathogens. A less appreciated, but equally interesting, PYHIN family member that forms an atypical inflammasome is IFI16 (interferon-inducible protein 16). IFI16 is comprised of a pyrin domain along with HIN-A and HIN-B domains [7]. IFI16 is present in humans but not in mice; although an ortholog, p204, is expressed in mice, its function remains unclear. Early studies of IFI16 determined that it is expressed in CD34+ myeloid precursor cells and also peripheral blood monocytes [8]. IFI16 is also expressed in T cells and epithelial cells from various tissues including the skin, gastrointestinal tracts, and urogenital tract [9, 10]. IFI16, unlike the cytosolic family member AIM2, is predominately present in the nucleus of the resting cells. Upon activation, IFI16 does however migrate from the nucleus to the cytoplasm. IFI16 is thought to be a nuclear pattern recognition receptor important for recognizing viruses that enter into the nucleus of cells. IFI16 recognizes a number of pathogens including Kaposi sarcoma-associated herpesvirus (KSHV). Infection of endothelial cells with KSHV led to the formation of an inflammasome complex containing IFI16, ASC, and caspase-1 in the nucleus and the cytoplasm of cells as seen in Fig. 1 [11]. IFI16 inflammasome activation by KSHV led to subsequent caspase-1 activation and IL-1β cleavage. IL-1β and IL-6 expressions along with caspase-1 activation in response to KSHV infection were dependent on expression of IFI16 and ASC indicating that IFI16, and not other inflammasomes, were being activated in response to KSHV infection. To further demonstrate that IFI16 is capable of forming

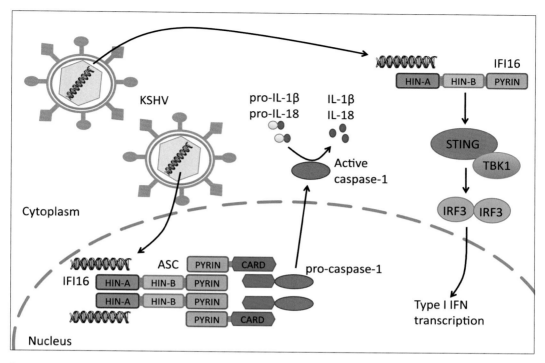

Fig. 1 Activation of IFI16 by viral DNA. In the cytoplasm, IFI16 recognizes DNA, either viral or synthetic, through the HIN domains. Upon recognition of DNA, IFI16 interacts with STING leading to TBK1-mediated phosphorylation of IRF3 and activation of NF-κB. IRF3 and NF-κB then translocate into the nucleus and induce transcription of IFN-α and IFN-β. In the nucleus of endothelial cells, IFI16 senses DNA from KSHV resulting in recruitment of the adaptor protein ASC and the cysteine protease caspase-1 to form an inflammasome. Active caspase-1 then cleaves pro-IL-1β and pro-IL-18 into their mature secreted forms. Upon activation, the IFI16 inflammasome migrates out of the nucleus into the cytoplasm

an inflammasome, 293T cells were transfected with pro-IL-1β, pro-caspase-1, ASC, and IFI-16 and then infected with KSHV. Cells transfected with inflammasome components secreted IL-1β indicating that IFI16 is indeed able to form a functional inflammasome [11]. The IFI16 inflammasome was activated by KSHV; however, other viruses such as herpes simplex virus 1 (HSV1) also enter into the nucleus of cells. It is reasonable to postulate that other inflammasomes may also assemble within the nucleus of cells in response to viral infection. The presence of both cytosolic and nuclear inflammasomes demonstrates the immune systems multi-faceted approach to detecting intracellular pathogens.

IFI16, in addition to forming an inflammasome, also regulates the type I IFN response during viral infection or in response to synthetic DNA [12]. IFI16 binds to viral DNA and mediates production of IFN-β. IFI16 directly interact with STING, which is known to be critical for IFN-β production in response to viral DNA [12]. Interaction between IFI16 and STING leads to TBK1-mediated phosphorylation of IRF3 and activation of NF-κB leading to transcription of IFN-α and IFN-β (Fig. 1) [12–14]. Production of IFN-β in response to DNA viruses was dependent on both IFI16 and STING therefore indicating that IFI16 modulates the STING pathway. IFI16 has proven to be a particularly interesting pattern recognition receptor with dual roles in the cytoplasm and nucleus. It is also evident that IFI16 is necessary for the host immune response to viral infection through production of type I interferons and IL-1β.

3 NLRP6: Regulator of Gut Microbiota

The NLR family of proteins is a large family of pattern recognition receptors with 23 defined members in humans and 34 in mice [1, 15]. NLR proteins have three conserved domains: a central nucleotide-binding and oligomerization domain (NACHT), a C-terminal leucine-rich repeat domain (LRR), and an N-terminal effector domain [16, 17]. The particular N-terminal effector domain further subgroups them; those that contain an N-terminal pyrin domain are members of the NLRP subgroup and those with a caspase activation and recruitment domain (CARD) are part of the NLRC subgroup. NLRP6 is expressed by a variety of hematopoietic cells including dendritic cells, neutrophils, macrophages, and T cells. NLRP6 is also highly expressed in the duodenum, ileum, and colon specifically in colonic epithelial cells [18, 19]. Due to the high expression of NLRP6 in the intestines, it is not surprising that NLRP6 plays an important role in intestinal homeostasis.

NLRP6, like many of the NLR proteins, has multiple roles in the immune response. In addition to being a regulator of intestinal homeostasis, NLRP6 also dampens inflammation and inflammatory signaling during bacterial infection [20]. Mice deficient in NLRP6 exhibited decreased mortality and bacterial burdens when challenged with *L. monocytogenes* and *Salmonella enterica* serovar Typhimurium. NLRP6-deficient mice also had reduced neutrophil influx in response to these pathogens [20]. Mechanistically, NLRP6-deficient macrophages had increased NF-κB activation and ERK signaling in response to bacterial infection [20]. Although dampening inflammation during bacterial infection may be detrimental to the host, NLRP6 may also help dampen aberrant immunopathology and damage to host tissues.

A central focus of NLRP6 research has been on the role NLRP6 plays in the intestines. NLRP6 has been heavily studied in models of colitis and colitis-associated colorectal cancer (CAC). Mice deficient in NLRP6 had increased tumor burden and pathology compared to wild-type mice in a model of azoxymethane and dextran sodium sulfate (AOM-DSS) CAC [19, 21]. NLRP6-deficient mice had increased expression of TNF-α, IL-1β, and

IFN-γ in the colonic tissue indicating an inability of NLRP6-deficient mice to control inflammation. NLRP6 was not only necessary for control of inflammation but also epithelial repair in the intestines [19, 21]. Taken together, these studies suggest that NLRP6 prevents excessive inflammation and epithelial damage, which decreases pathology and susceptibility to CAC.

Studies focusing on the microbial ecology in the intestines of mice revealed the NLRP6 inflammasome as master regulator of the intestinal microbiota [22]. NLRP6 was previously described to form an inflammasome in vitro. Expression of NLRP6 and ASC in 293T cells resulted in the recruitment of NLRP6 to ASC speck-like structures. In addition, COS-7L cells transfected with plasmids encoding pro-caspase-1, ASC, and NLRP6 were shown to secrete IL-1β [18]. NLRP6 inflammasome formation was further suggested by the observation that similar to NLRP6-deficient mice, ASC- and IL-18-deficient mice also developed enhanced DSS-induced colitis. Interestingly, cohousing of wild-type mice with NLRP6- or ASC-deficient mice led to enhanced disease in the wild-type mice [22]. This identified the microbiota as being a driving factor of enhanced colitis. The NLRP6-deficient microbiota had enhanced phylum Bacteroidetes family *Prevotellaceae*. Outgrowth of *Prevotellaceae* resulted in enhanced colonic inflammation and increased levels of the chemokine CCL5. CCL5 is an important neutrophil chemoattractant and may contribute to neutrophil-mediated pathology in the intestines. In these studies, IL-18 production was also key in preserving intestinal homeostasis [22]. In the absence of IL-18, mice developed significantly enhanced disease compared to wild-type mice. It is hypothesized that the NLRP6 inflammasome is activated in the intestinal epithelial cells by an unknown ligand. Activation of the inflammasome leads to processing of IL-18 into its mature form, which then through an unknown mechanism regulates microbiota to prevent dysbiosis (Fig. 2) [22].

Excessive inflammation in the intestines is commonly caused by a breakdown of the intestinal barrier. An important component of the intestinal barrier is a thick layer of mucus that separates the intestinal epithelium from microbes present in the intestines. Further insight into the role of the NLRP6 inflammasome in intestinal homeostasis was provided by studies looking at mucus production in the intestines of mice challenged with the enteric pathogen *Citrobacter rodentium* [23]. Mice deficient in NLRP6, ASC, or caspase-1 all had a marked decrease in mucus thickness leading to increased cell adherence by *C. rodentium* followed by increased bacterial dissemination. The decrease in mucus thickness was found to be due to a deficiency in mucin granule exocytosis in goblet cells. Mechanistically, it was determined that defects in autophagy led to altered goblet cell function and mucus secretion [23]. It is still unclear what the ligand that activates the NLRP6

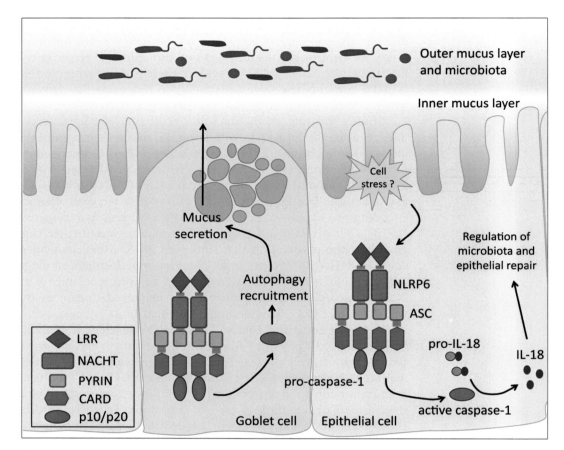

Fig. 2 NLRP6 inflammasome-mediated regulation of the gut microbiota. NLRP6 is highly expressed in the intestines of mice, and its expression is required for maintenance of the intestinal microbiota. Upon activation by an unknown ligand, NLRP6 associates with ASC and caspase-1 to form an inflammasome. Inflammasome activation results in cleavage and secretion of IL-18. IL-18 production is necessary to prevent outgrowth of the colitogenic bacteria *Prevotellaceae* in the intestines and to prevent tissue damage caused by the immune response to *Prevotellaceae*. The NLRP6 inflammasome also plays a role in goblet cell function. Through an undefined mechanism, the NLRP6 inflammasome mediates mucin granule exocytosis in goblet cells and is important for maintaining the mucus barrier in the intestines

inflammasome is in the intestines. However, it is clear that the NLRP6 inflammasome helps provide a protective intestinal barrier and regulate dysbiosis (Fig. 2).

4 NLRP7: Regulator of Embryonic Development and Inflammation

The NLR family of proteins has a wide array of functions outside of pathogen and cellular damage recognition. NLRP7 is a prime example of this. NLRP7 has multiple roles including regulating inflammation and recognition of microbial lipopeptides and plays a critical role in embryonic development [24–26]. The structure of NLRP7 includes an N-terminal pyrin domain, which is involved in protein-protein interactions and downstream signaling pathways, a central large NACHT domain that contains a nuclear localization signal and an LRR domain that varies in length depending on splicing [27]. NLRP7 is only expressed in humans and the closest mouse analog is NLRP2, which is also involved in embryonic development [28]. NLRP7 is expressed in human cell lines including B cell lines, T cell lines, and monocytic cell lines [28]. NLRP7 is also expressed in human lung, spleen, thymus, and reproductive organs including the testis and ovaries [25, 29]. In particular, NLRP7 is expressed in oocytes in the ovaries, and this expression highlights the important role it plays in embryonic development [30].

NLRP7 research is heavily focused on the role it plays in recurrent hydatidiform mole formation. Hydatidiform moles are a gestational trophoblastic disease with an incidence of approximately 1 in 600 pregnancies in Western countries with a higher incidence rate in less-developed countries [31, 32]. A hydatidiform mole is an aberrant pregnancy that is characterized by a mass exhibiting trophoblastic hyperplasia and swelling of chorionic villi resulting in a nonviable pregnancy [33]. Hydatidiform moles can be characterized as either a complete mole or partial mole. Through genetic screening, NLRP7 was identified as the causative gene [26]. Subsequently, a wide range of NLRP7 mutations have been identified in patients, including mutations resulting in a truncated protein and also missense mutations, among others [26, 34, 35]. A number of these mutations were identified in the LRR region, which is proposed to be important for ligand sensing [30]. The mechanism by which NLRP7 contributes to embryonic development is still unclear. NLRP7 may be mediating cytokine production in oocytes that is necessary for development. However, the ligand being sensed by NLRP7, or proteins that NLRP7 is interacting with, in the oocyte has not been identified. It is becoming increasingly clear that the NLR family is critical for embryonic development as family members NLRP7, NLRP2, NLRP5, and NLRP9 all contribute to this process [36].

Early in vitro studies involving NLRP7 (also known as PYPAF3) demonstrated a role for NLRP7 as a negative regulator of inflammation. In transiently transfected HEK293 cells, NLRP7 directly interacted with pro-caspase-1 and pro- IL-1β and inhibited IL-1β processing into its mature form [25]. The same inhibition was also observed in THP-1 cells. Upon stimulation with lipopolysaccharide (LPS) or IL-1β, NLRP7 was upregulated in THP-1 cells. Furthermore, THP-1 cells stably expressing NLRP7 secreted less IL-1β than cells transfected with an empty vector when treated with LPS [25]. The role of NLRP7 as a negative regulator of IL-1β processing was confirmed using similar transient transfection experiments [28]. In these studies, the pyrin domain was identified as the critical domain for inhibition of IL-1β processing. In the same study, NLRP7 was shown to co-localize with the microtubule-organizing center in peripheral blood mononuclear cells [28]. It was postulated that this co-localization is important for cytokine trafficking out of the cell. It should be noted that peripheral blood mononuclear cells isolated from patients with NLRP7 mutations secrete low levels of IL-1β in response to LPS challenge compared to controls [28]. This would suggest NLRP7 is important for cleavage of IL-1β. However, when NLRP7 is present in high levels it inhibits IL-1β cleavage, hence the role of NLRP7 may be dependent of levels of expression.

NLRP7 was recently shown to interact with ASC and caspase-1 to form an inflammasome in response to *Mycoplasma* spp. [24]. NLRP7 was identified through the use of a siRNA knockdown screen of various NLRs in THP-1 cells treated with heat-killed *Acholeplasma laidlawii*. THP-1 cells transfected with NLRP7 siRNA exhibited a marked decrease in IL-1β production in response to heat-killed *A. laidlawii*, *Staphylococcus aureus*, and *L. monocytogenes* [24]. However, NLRP7 expression did not alter cell death suggesting that NLRP7 inflammasome formation does not induce pyroptosis. NLRP7 also triggered caspase-1-mediated IL-1β production in response to various bacterial acylated lipopeptides and known TLR2 agonists including FSL-1, MALP-2, Pam2CSK4, and Pam3CSK4 (Fig. 3). Furthermore, activation of TLR2 was required for NLRP7-mediated IL-1β maturation. TLR2 activation in response to acylated lipopeptides was needed for transcription of both pro-IL-1β and pro-IL-18 and therefore serves as a potential priming step for NLRP7 inflammasome activation. These data demonstrates a novel role for NLRP7 in recognition of microbial lipopeptides [24]. Whether NLRP7 forms an inflammasome in other cell types, such as oocytes, and whether other ligands exist have yet to be determined and present an interesting area for future research.

5   NLRP12: Regulator of Inflammation

The structure of NLRP12 includes an N-terminal pyrin domain, a NACHT domain, and a C-terminal LRR [37]. NLRP12 is highly expressed in hematopoietic cells including eosinophils, neutrophils, macrophages, and dendritic cells [38, 39]. NLRP12 has been shown to modulate NF-κB activation in in vitro transient transfection systems; however, the precise role of NLRP12 in activation of NF-κB remains unclear. Transient transfection of HEK293T cells with NLRP12 and ASC constructs leads to transcription of an NF-κB luciferase reporter plasmid [38]. In contrast, NLRP12 was shown to suppress noncanonical NF-κB activation in THP-1 cells. NLRP12 suppressed noncanonical NF-κB activation

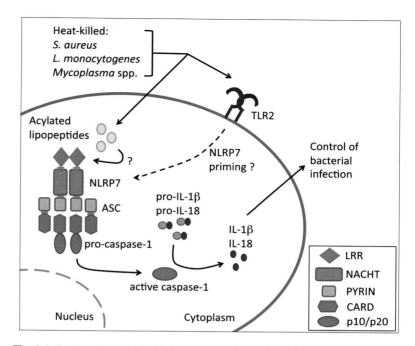

Fig. 3 Activation of the NLRP7 inflammasome by acylated lipopeptides. Challenge with *Acholeplasma laidlawii*, *Staphylococcus aureus*, and *Listeria monocytogenes* triggers NLRP7 inflammasome mediated cleavage of IL-1β and IL-18. More specifically, acylated lipopeptides and known TLR2 agonists including FSL-1, MALP-2, Pam2CSK4, and Pam3CSK4 activate the NLRP7 inflammasome. TLR2 recognition of acylated lipopeptides is required for activation of the NLRP7 inflammasome suggesting TLR2 activation may serve as an important priming step

by binding to NF-κB-inducing kinase (NIK) leading to the degradation of NIK [40].

NLRP12 acting as a negative regulator of noncanonical NF-κB activation was also seen in vivo [41]. In a mouse model of CAC, *Nlrp12*-deficient mice exhibited increased pathology and tumor burdens. *Nlrp12*-deficient mice expressed elevated levels of the chemokines CXCL12 and CXCL13 and also had increased phosphorylation of ERK compared to wild-type mice [41]. It was hypothesized that NLRP12 negatively regulates noncanonical NF-κB signaling via interactions with NIK and TRAF3 (Fig. 4). This coincides with previous in vitro studies that showed NLRP12 binding to NIK and suppressing noncanonical NF-κB activation [40, 41]. By binding to TRAF3 and NIK, NLRP12 is able to act as a negative regulator of inflammation and dampen cancer-promoting chemokines in vivo.

It is well documented that NLRP12 plays a protective role in mouse models of CAC. In a different study, NLRP12-deficient mice also exhibited an increase in tumor numbers compared to wild-type mice utilizing the same AOM-DSS model of CAC [42].

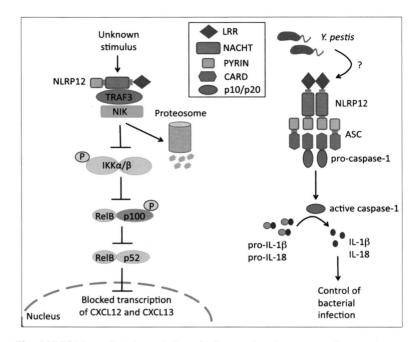

Fig. 4 NLRP12-mediated regulation of inflammation. In response to an unknown ligand, NLRP12 suppresses noncanonical NF-κB activation. NLRP12 binds to NF-κB-inducing kinase (NIK) and TRAF3 leading to the degradation of NIK and further blocks downstream signaling events. The decrease in noncanonical NF-κB signaling leads to decreased expression of chemokines CXCL12 and CXCL13 and reduced inflammation in mouse models of colorectal cancer. In response to *Yersinia pestis*, NLRP12 associates with ASC and caspase-1 to form an inflammasome. IL-18-mediated cleavage by the NLRP12 inflammasome is necessary for control of bacterial replication and survival in mice

However, a different mechanism of tumor modulation was proposed. Colons from *Nlrp12*-deficient mice displayed increased tissue damage, inflammatory cell infiltrate, and pro-inflammatory cytokine production. Increased activation of signaling pathways including ERK, STAT3, and NF-κB were also observed in *Nlrp12*-deficient mice [42]. More specifically, expression of NLRP12 in hematopoietic cells was critical for ERK and NF-κB signaling and subsequent tumor protection. Further in vitro analysis confirmed increased phosphorylation of ERK and IκBα in *Nlrp12*-deficient bone marrow-derived macrophages (BMDMs) when stimulated with TLR ligands [42]. In this specific study, NLRP12 was shown to modulate the canonical NF-κB pathway in BMDMs [42]. Thus, it was concluded that dysregulation of canonical NF-κB signaling in *Nlrp12*-deficient hematopoietic cells, most likely macrophages, leads to increased tumor-driving inflammation. It is still unclear whether NLRP12 modulates the canonical or noncanonical NF-κB pathway. Differences seen may be due to mouse background differences, differences in animal housing, or differences in the model used. It is clear, however, that NLRP12 is an important regulator of colonic carcinogenesis.

Early in vitro studies demonstrated that NLRP12 is capable of interacting with ASC in HEK293T cells [38]. In the same study, NLRP12 was shown to activate caspase-1 in COS-7L cells transfected with ASC, pro-caspase-1, and NLRP12 plasmids. Although extensive biochemical studies have not been performed, it is hypothesized that NLRP12 forms an inflammasome with ASC and caspase-1. Activation of the NLRP12 inflammasome is crucial for control of *Yersinia pestis* infection [43]. Mice deficient in NLRP12 had reduced survival compared to wild type when infected with an attenuated strain of *Y. pestis*. It was further shown that NLRP12 was necessary for caspase-1-mediated cleavage of IL-18 [43]. Subsequently, IL-18 was necessary for control of bacterial infection by mediating production of IFN-γ (Fig. 4).

Recently, the role of NLRP12 during bacterial infection has been further explored. Although NLRP12 was important for control of *Y. pestis* infection, it did not play a role during infection with *Klebsiella pneumoniae* and *Mycobacterium tuberculosis* [44]. In contrast, NLRP12 expression had deleterious effects during *S. typhimurium* infection [45]. Mice deficient in NLRP12 were resistant to *S. typhimurium* and exhibited reduced bacterial burdens. The decrease in bacterial burdens was attributed to an increase in pro-inflammatory cytokines and increased activation of ERK and NF-κB pathways [45]. These data are consistent with previous studies and demonstrate NLRP12 functions as a negative regulator of inflammation. Regulating the ERK and NF-κB pathways is important for controlling inflammation in the context of cancer but is detrimental during bacterial infection when inflammation is required for bacterial clearance. Importantly, however, in both of these studies, no defect in caspase-1-mediated IL-1β was observed in the absence of NLRP12. Together, studies on NLRP12 suggest that it functions as a negative regulator of inflammatory signaling. Future studies will be required to determine the ligand that activates NLRP12 and to determine if NLRP12 truly forms a functional inflammasome complex.

6 NLRC5: A Critical Regulator of MHCI Expression

NLRC5 belongs to the NLRC subfamily of proteins and possesses an N-terminal CARD, a large central NACHT domain, and a C-terminal LRR domain. The most well-characterized NLRC proteins include NLRC1 (NOD1), NLRC2 (NOD2), and NLRC4. NOD1 and NOD2 recognize the bacterial peptidoglycan fragments from both Gram-positive and Gram-negative bacteria [46–48]. The inflammasome-forming protein NLRC4 is critical for recognition of bacterial flagellin and parts of the type III and IV secretion systems [49–51]. Together the NLRC family makes up an important component of the innate defense against bacterial pathogens. NLRC5, like its fellow family members, is also important in antibacterial defenses; however, it also has an interesting role outside of pathogen recognition as a regulator of MHC class I expression.

Structurally, NLRC5 has the longest LRR domain in the NLR family with 27 LRRs making it almost double the size of its other family members [52]. Additionally, the CARD domain of NLRC5 is not homologous to the other NLRC proteins. NLRC5 most closely resembles the NLR family member class II transactivator (CIITA) [52]. CIITA is pivotal for expression of MHC class II genes and additionally has a role in MHC class I gene transcription [53, 54]. Both NLRC5 and CIITA have N-terminal nuclear localization signals, and NLRC5 is found in both the cytoplasm and nucleus of cells [55]. The homology between CIITA and NLRC5 may explain their similar roles in regulation of MHC gene expression. NLRC5 is expressed in a wide array of cell types including immune cells such as macrophages, dendritic cells, T cells, B cells, and fibroblasts [56]. Expression of NLRC5 is induced by IFN-γ, which leads to STAT1-mediated gene transcription [56].

NLRC5, as mentioned above, can function as a regulator of MHC class I gene expression. Presentation of intracellular antigens on MHC class I molecules is imperative for mounting appropriate CD8 T cell response. Subsequently, without a CD8 T cell response, the host is left susceptible to viral and bacterial infections. NLRC5 also regulates MHC class I-associated genes including β_2-microglobulin, Tap1, and Lmp2 through interaction with the MHC class I enhanceosome [55, 57–59]. In the absence of NLRC5, either through gene knockdown or in knockout mice, an overall decrease in the surface expression of MHCI class I in a wide array of cell types was observed [57, 58]. Structurally, the NACHT domain and the nuclear localization signal of NLRC5 are required for nuclear localization and transcription of MHC class I [60]. Lack of expression of MHC I genes in NLRC5-deficient mice also directly impacted the CD8 T cell response during *L. monocytogenes* infection [57]. NLRC5-deficient mice exhibited increased bacterial burdens due to decreased IFN-γ production by antigen-specific CD8 T cells [57].

NLRC5 is expressed both in the nucleus and cytoplasm of cells. Although the role of NLRC5 within the nucleus as a regulator of gene expression is widely accepted, the role of NLRC5 within the cytoplasm is less clear. NLRC5 has been described to be a negative regulator of NF-κB and type I interferon signaling [61–63]. In contrast, during viral challenge, NLRC5 has been shown to interact with RIG-I and positively regulates type I

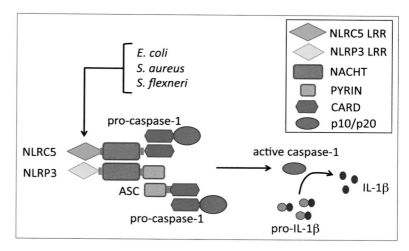

Fig. 5 NLRC5 and NLRP3 cooperative inflammasome complex. In response to infection with bacteria such as *Escherichia coli, Staphylococcus aureus*, and *Shigella flexneri*, NLRC5 interacts with NLRP3 through the NACHT domain. NLRP3 recruits ASC and caspase-1 to the complex and mediates cleavage of IL-1β. The presence of both NLRC5 and NLRP3 has a synergistic effect, and enhanced levels of mature IL-1β are produced. Therefore, it is possible that NLRC5 interacts with the NLRP3 inflammasome to positively modulate its activation. NLRC5 is also capable of activating caspase-1 and cleaving IL-1β in the absence of NLRP3. Therefore, NLRC5 through its CARD domain may also recruit caspase-1 directly to the NLRC5/NLRP3 inflammasome complex

interferon signaling [64, 65]. Differences in cell type, level of NLRC5 expression, and type of stimulus used to challenge cells could possibly be responsible for the variation in results. Thus, the precise role of NLRC5 in modulation of innate immune signaling pathways may be variable depending on the system used.

Another described function of NLRC5 is in the formation of an inflammasome. Knockdown of NLRC5 in monocytes resulted in decreased IL-1β secretion and caspase-1 activation in response to infection with *Escherichia coli* [56]. Similar decreases in IL-1β secretion were seen upon infection with *S. aureus* and *Shigella flexneri* or stimulation with TLR ligands [56]. Interestingly, NLRC5 modulated IL-1β secretion in response to known NLRP3 agonists. NLRC5 was also found to physically interact with NLRP3 and ASC (Fig. 5). Binding of NLRC5 to NLRP3 required an intact NACHT domain [56]. Expression of ASC, pro-IL-1β, pro-caspase-1, NLRC5, and NLRP3 in HEK293T cells led to cleavage of IL-1β, and the expression of NLRC5 and NLRP3 together appeared to have a synergistic effect on IL-1β cleavage. Therefore, it is possible that NLRC5 interacts with the NLRP3 inflammasome to positively modulate its activation.

It should be noted that conflicting results were seen with experiments utilizing bone marrow-derived dendritic cells (BMDCs) from NLRC5-deficient mice. There was no difference in cytokine production, including IL-1β, between wild-type and NLRC5-deficient BMDCs challenged with various TLR ligands and known NLRP3 agonists [66]. However, when HEK293 cells were reconstituted with NLRC5, pro-IL-1β, and pro-caspase-1, IL-1β was cleaved into its mature form [66]. These data are partially consistent with the above studies [56] and suggest that NLRC5 is capable of forming a functional inflammasome. Additional studies are still required to determine if NLRC5 is capable of independently forming an inflammasome complex and to further elucidate the biological consequences of modulating NLRP3 inflammasome function.

7   Concluding Remarks

The NLR and PYHIN family of proteins have a diverse array of functions in addition to the recognition of pathogens and cellular damage. Members of the NLR and PYHIN families have essential roles in embryonic development, regulation of key inflammatory signaling pathways, expression of MHC genes, and inflammasome formation. How the various functions of these pattern recognition receptors (PRRs) are regulated is still ambiguous. It is possible that the function of various PRRs is ligand dependent. A specific ligand may trigger inflammasome formation, while recognition of another ligand may lead to regulation of signaling pathways such as NF-κB. Ligand recognition may also impact conformational changes of the PRR, which in turn may impact function. Identifying the different ligands that these proteins recognize will be a critical step in determining how they function. Additionally, it is likely that the role of PRRs is also dependent on tissue and cell type. Within the NLR and PYHIN families, there are PRRs with similar functions including atypical and typical inflammasomes. A study of PRRs outside of the four classic inflammasomes provides great insight into the multiple functions that these family of molecules have and the importance they play in the immune response.

Acknowledgments

NIH grants R01 AI087630 (F.S.S.) and T32 AI007485 (A.M.J.) supported this work.

References

1. Schroder K, Tschopp J (2010) The inflammasomes. Cell 140:821–832. doi:10.1016/j.cell.2010.01.040
2. Schattgen SA, Fitzgerald KA (2011) The PYHIN protein family as mediators of host defenses. Immunol Rev 243(1):109–118. doi:10.1111/j.1600-065X.2011.01053.x
3. Fernandes-Alnemri T, Yu JW, Datta P, Wu J, Alnemri ES (2009) AIM2 activates the inflammasome and cell death in response to cytoplasmic DNA. Nature 458(7237):509–513. doi:10.1038/nature07710
4. Fernandes-Alnemri T, Yu JW, Juliana C, Solorzano L, Kang S, Wu J, Datta P, McCormick M, Huang L, McDermott E, Eisenlohr L, Landel CP, Alnemri ES (2010) The AIM2 inflammasome is critical for innate immunity to Francisella tularensis. Nat Immunol 11(5):385–393. doi:10.1038/ni.1859
5. Jones JW, Kayagaki N, Broz P, Henry T, Newton K, O'Rourke K, Chan S, Dong J, Qu Y, Roose-Girma M, Dixit VM, Monack DM (2010) Absent in melanoma 2 is required for innate immune recognition of Francisella tularensis. Proc Natl Acad Sci U S A 107(21):9771–9776. doi:10.1073/pnas.1003738107
6. Rathinam VA, Jiang Z, Waggoner SN, Sharma S, Cole LE, Waggoner L, Vanaja SK, Monks BG, Ganesan S, Latz E, Hornung V, Vogel SN, Szomolanyi-Tsuda E, Fitzgerald KA (2010) The AIM2 inflammasome is essential for host defense against cytosolic bacteria and DNA viruses. Nat Immunol 11(5):395–402. doi:10.1038/ni.1864
7. Ludlow LE, Johnstone RW, Clarke CJ (2005) The HIN-200 family: more than interferon-inducible genes? Exp Cell Res 308(1):1–17. doi:10.1016/j.yexcr.2005.03.032
8. Dawson MJ, Elwood NJ, Johnstone RW, Trapani JA (1998) The IFN-inducible nucleoprotein IFI 16 is expressed in cells of the monocyte lineage, but is rapidly and markedly down-regulated in other myeloid precursor populations. J Leukoc Biol 64(4):546–554
9. Gariglio M, Azzimonti B, Pagano M, Palestro G, De Andrea M, Valente G, Voglino G, Navino L, Landolfo S (2002) Immunohistochemical expression analysis of the human interferon-inducible gene IFI16, a member of the HIN200 family, not restricted to hematopoietic cells. J Interferon Cytokine Res 22(7):815–821. doi:10.1089/107999002320271413
10. Wei W, Clarke CJ, Somers GR, Cresswell KS, Loveland KA, Trapani JA, Johnstone RW (2003) Expression of IFI 16 in epithelial cells and lymphoid tissues. Histochem Cell Biol 119(1):45–54. doi:10.1007/s00418-002-0485-0
11. Kerur N, Veettil MV, Sharma-Walia N, Bottero V, Sadagopan S, Otageri P, Chandran B (2011) IFI16 acts as a nuclear pathogen sensor to induce the inflammasome in response to Kaposi Sarcoma-associated herpesvirus infection. Cell Host Microbe 9(5):363–375. doi:10.1016/j.chom.2011.04.008
12. Unterholzner L, Keating SE, Baran M, Horan KA, Jensen SB, Sharma S, Sirois CM, Jin T, Latz E, Xiao TS, Fitzgerald KA, Paludan SR, Bowie AG (2010) IFI16 is an innate immune sensor for intracellular DNA. Nat Immunol 11(11):997–1004. doi:10.1038/ni.1932
13. Ishikawa H, Ma Z, Barber GN (2009) STING regulates intracellular DNA-mediated, type I interferon-dependent innate immunity. Nature 461(7265):788–792. doi:10.1038/nature08476
14. Stetson DB, Medzhitov R (2006) Recognition of cytosolic DNA activates an IRF3-dependent innate immune response. Immunity 24(1):93–103. doi:10.1016/j.immuni.2005.12.003
15. Ting JP, Davis BK (2005) CATERPILLER: a novel gene family important in immunity, cell death, and diseases. Annu Rev Immunol 23:387–414. doi:10.1146/annurev.immunol.23.021704.115616
16. Martinon F, Mayor A, Tschopp J (2009) The inflammasomes: guardians of the body. Annu Rev Immunol 27:229–265. doi:10.1146/annurev.immunol.021908.132715
17. Ye Z, Ting JP (2008) NLR, the nucleotide-binding domain leucine-rich repeat containing gene family. Curr Opin Immunol 20(1):3–9. doi:10.1016/j.coi.2008.01.003
18. Grenier JM, Wang L, Manji GA, Huang WJ, Al-Garawi A, Kelly R, Carlson A, Merriam S, Lora JM, Briskin M, DiStefano PS, Bertin J (2002) Functional screening of five PYPAF family members identifies PYPAF5 as a novel regulator of NF-kappaB and caspase-1. FEBS Lett 530(1–3):73–78
19. Normand S, Delanoye-Crespin A, Bressenot A, Huot L, Grandjean T, Peyrin-Biroulet L, Lemoine Y, Hot D, Chamaillard M (2011) Nod-like receptor pyrin domain-containing protein 6 (NLRP6) controls epithelial self-renewal and colorectal carcinogenesis upon injury. Proc Natl Acad Sci U S A 108(23):9601–9606. doi:10.1073/pnas.1100981108

20. Anand PK, Malireddi RK, Lukens JR, Vogel P, Bertin J, Lamkanfi M, Kanneganti TD (2012) NLRP6 negatively regulates innate immunity and host defence against bacterial pathogens. Nature 488(7411):389–393. doi:10.1038/nature11250

21. Chen GY, Liu M, Wang F, Bertin J, Nunez G (2011) A functional role for Nlrp6 in intestinal inflammation and tumorigenesis. J Immunol 186(12):7187–7194. doi:10.4049/jimmunol.1100412

22. Elinav E, Strowig T, Kau AL, Henao-Mejia J, Thaiss CA, Booth CJ, Peaper DR, Bertin J, Eisenbarth SC, Gordon JI, Flavell RA (2011) NLRP6 inflammasome regulates colonic microbial ecology and risk for colitis. Cell 145(5):745–757. doi:10.1016/j.cell.2011.04.022

23. Wlodarska M, Thaiss CA, Nowarski R, Henao-Mejia J, Zhang JP, Brown EM, Frankel G, Levy M, Katz MN, Philbrick WM, Elinav E, Finlay BB, Flavell RA (2014) NLRP6 inflammasome orchestrates the colonic host-microbial interface by regulating goblet cell mucus secretion. Cell 156(5):1045–1059. doi:10.1016/j.cell.2014.01.026

24. Khare S, Dorfleutner A, Bryan NB, Yun C, Radian AD, de Almeida L, Rojanasakul Y, Stehlik C (2012) An NLRP7-containing inflammasome mediates recognition of microbial lipopeptides in human macrophages. Immunity 36(3):464–476. doi:10.1016/j.immuni.2012.02.001

25. Kinoshita T, Wang Y, Hasegawa M, Imamura R, Suda T (2005) PYPAF3, a PYRIN-containing APAF-1-like protein, is a feedback regulator of caspase-1-dependent interleukin-1beta secretion. J Biol Chem 280(23):21720–21725. doi:10.1074/jbc.M410057200

26. Murdoch S, Djuric U, Mazhar B, Seoud M, Khan R, Kuick R, Bagga R, Kircheisen R, Ao A, Ratti B, Hanash S, Rouleau GA, Slim R (2006) Mutations in NALP7 cause recurrent hydatidiform moles and reproductive wastage in humans. Nat Genet 38(3):300–302. doi:10.1038/ng1740

27. Tschopp J, Martinon F, Burns K (2003) NALPs: a novel protein family involved in inflammation. Nat Rev Mol Cell Biol 4(2):95–104. doi:10.1038/nrm1019

28. Messaed C, Akoury E, Djuric U, Zeng J, Saleh M, Gilbert L, Seoud M, Qureshi S, Slim R (2011) NLRP7, a nucleotide oligomerization domain-like receptor protein, is required for normal cytokine secretion and co-localizes with Golgi and the microtubule-organizing center. J Biol Chem 286(50):43313–43323. doi:10.1074/jbc.M111.306191

29. Okada K, Hirota E, Mizutani Y, Fujioka T, Shuin T, Miki T, Nakamura Y, Katagiri T (2004) Oncogenic role of NALP7 in testicular seminomas. Cancer Sci 95(12):949–954

30. Wang CM, Dixon PH, Decordova S, Hodges MD, Sebire NJ, Ozalp S, Fallahian M, Sensi A, Ashrafi F, Repiska V, Zhao J, Xiang Y, Savage PM, Seckl MJ, Fisher RA (2009) Identification of 13 novel NLRP7 mutations in 20 families with recurrent hydatidiform mole; missense mutations cluster in the leucine-rich region. J Med Genet 46(8):569–575. doi:10.1136/jmg.2008.064196

31. Grimes DA (1984) Epidemiology of gestational trophoblastic disease. Am J Obstet Gynecol 150(3):309–318

32. Savage P, Williams J, Wong SL, Short D, Casalboni S, Catalano K, Seckl M (2010) The demographics of molar pregnancies in England and Wales from 2000–2009. J Reprod Med 55(7–8):341–345

33. Berkowitz RS, Goldstein DP (2009) Clinical practice. Molar pregnancy. N Engl J Med 360(16):1639–1645. doi:10.1056/NEJMcp0900696

34. Kou YC, Shao L, Peng HH, Rosetta R, del Gaudio D, Wagner AF, Al-Hussaini TK, Van den Veyver IB (2008) A recurrent intragenic genomic duplication, other novel mutations in NLRP7 and imprinting defects in recurrent biparental hydatidiform moles. Mol Hum Reprod 14(1):33–40. doi:10.1093/molehr/gam079

35. Qian J, Deveault C, Bagga R, Xie X, Slim R (2007) Women heterozygous for NALP7/NLRP7 mutations are at risk for reproductive wastage: report of two novel mutations. Hum Mutat 28(7):741. doi:10.1002/humu.9498

36. Van Gorp H, Kuchmiy A, Van Hauwermeiren F, Lamkanfi M (2014) NOD-like receptors interfacing the immune and reproductive systems. FEBS J 281(20):4568–4582. doi:10.1111/febs.13014

37. Pinheiro AS, Eibl C, Ekman-Vural Z, Schwarzenbacher R, Peti W (2011) The NLRP12 pyrin domain: structure, dynamics, and functional insights. J Mol Biol 413(4):790–803. doi:10.1016/j.jmb.2011.09.024

38. Wang L, Manji GA, Grenier JM, Al-Garawi A, Merriam S, Lora JM, Geddes BJ, Briskin M, DiStefano PS, Bertin J (2002) PYPAF7, a novel PYRIN-containing Apaf1-like protein that regulates activation of NF-kappa B and caspase-1-dependent cytokine processing. J Biol Chem 277(33):29874–29880. doi:10.1074/jbc.M203915200

39. Williams KL, Lich JD, Duncan JA, Reed W, Rallabhandi P, Moore C, Kurtz S, Coffield VM, Accavitti-Loper MA, Su L, Vogel SN, Braunstein M, Ting JP (2005) The CATERPILLER protein monarch-1 is an

antagonist of toll-like receptor-, tumor necrosis factor alpha-, and Mycobacterium tuberculosis-induced pro-inflammatory signals. J Biol Chem 280(48):39914–39924. doi:10.1074/jbc.M502820200

40. Lich JD, Williams KL, Moore CB, Arthur JC, Davis BK, Taxman DJ, Ting JP (2007) Monarch-1 suppresses non-canonical NF-kappaB activation and p52-dependent chemokine expression in monocytes. J Immunol 178(3):1256–1260

41. Allen IC, Wilson JE, Schneider M, Lich JD, Roberts RA, Arthur JC, Woodford RM, Davis BK, Uronis JM, Herfarth HH, Jobin C, Rogers AB, Ting JP (2012) NLRP12 suppresses colon inflammation and tumorigenesis through the negative regulation of noncanonical NF-kappaB signaling. Immunity 36(5):742–754. doi:10.1016/j.immuni.2012.03.012

42. Zaki MH, Vogel P, Malireddi RK, Body-Malapel M, Anand PK, Bertin J, Green DR, Lamkanfi M, Kanneganti TD (2011) The NOD-like receptor NLRP12 attenuates colon inflammation and tumorigenesis. Cancer Cell 20(5):649–660. doi:10.1016/j.ccr.2011.10.022

43. Vladimer GI, Weng D, Paquette SW, Vanaja SK, Rathinam VA, Aune MH, Conlon JE, Burbage JJ, Proulx MK, Liu Q, Reed G, Mecsas JC, Iwakura Y, Bertin J, Goguen JD, Fitzgerald KA, Lien E (2012) The NLRP12 inflammasome recognizes Yersinia pestis. Immunity 37(1):96–107. doi:10.1016/j.immuni.2012.07.006

44. Allen IC, McElvania-Tekippe E, Wilson JE, Lich JD, Arthur JC, Sullivan JT, Braunstein M, Ting JP (2013) Characterization of NLRP12 during the in vivo host immune response to Klebsiella pneumoniae and Mycobacterium tuberculosis. PLoS One 8(4), e60842. doi:10.1371/journal.pone.0060842

45. Zaki MH, Man SM, Vogel P, Lamkanfi M, Kanneganti TD (2014) Salmonella exploits NLRP12-dependent innate immune signaling to suppress host defenses during infection. Proc Natl Acad Sci U S A 111(1):385–390. doi:10.1073/pnas.1317643111

46. Girardin SE, Boneca IG, Carneiro LA, Antignac A, Jehanno M, Viala J, Tedin K, Taha MK, Labigne A, Zahringer U, Coyle AJ, DiStefano PS, Bertin J, Sansonetti PJ, Philpott DJ (2003) Nod1 detects a unique muropeptide from gram-negative bacterial peptidoglycan. Science 300(5625):1584–1587. doi:10.1126/science.1084677

47. Girardin SE, Boneca IG, Viala J, Chamaillard M, Labigne A, Thomas G, Philpott DJ, Sansonetti PJ (2003) Nod2 is a general sensor of peptidoglycan through muramyl dipeptide (MDP) detection. J Biol Chem 278(11):8869–8872. doi:10.1074/jbc.C200651200

48. Strober W, Murray PJ, Kitani A, Watanabe T (2006) Signalling pathways and molecular interactions of NOD1 and NOD2. Nat Rev Immunol 6(1):9–20. doi:10.1038/nri1747

49. Miao EA, Mao DP, Yudkovsky N, Bonneau R, Lorang CG, Warren SE, Leaf IA, Aderem A (2010) Innate immune detection of the type III secretion apparatus through the NLRC4 inflammasome. Proc Natl Acad Sci U S A 107(7):3076–3080. doi:10.1073/pnas.0913087107

50. Sutterwala FS, Mijares LA, Li L, Ogura Y, Kazmierczak BI, Flavell RA (2007) Immune recognition of Pseudomonas aeruginosa mediated by the IPAF/NLRC4 inflammasome. J Exp Med 204(13):3235–3245. doi:10.1084/jem.20071239

51. Zhao Y, Yang J, Shi J, Gong YN, Lu Q, Xu H, Liu L, Shao F (2011) The NLRC4 inflammasome receptors for bacterial flagellin and type III secretion apparatus. Nature 477(7366):596–600. doi:10.1038/nature10510

52. Yao Y, Qian Y (2013) Expression regulation and function of NLRC5. Protein Cell 4(3):168–175. doi:10.1007/s13238-012-2109-3

53. Martin BK, Chin KC, Olsen JC, Skinner CA, Dey A, Ozato K, Ting JP (1997) Induction of MHC class I expression by the MHC class II transactivator CIITA. Immunity 6(5):591–600

54. Steimle V, Siegrist CA, Mottet A, Lisowska-Grospierre B, Mach B (1994) Regulation of MHC class II expression by interferon-gamma mediated by the transactivator gene CIITA. Science 265(5168):106–109

55. Kobayashi KS, van den Elsen PJ (2012) NLRC5: a key regulator of MHC class I-dependent immune responses. Nat Rev Immunol 12(12):813–820. doi:10.1038/nri3339

56. Davis BK, Roberts RA, Huang MT, Willingham SB, Conti BJ, Brickey WJ, Barker BR, Kwan M, Taxman DJ, Accavitti-Loper MA, Duncan JA, Ting JP (2011) Cutting edge: NLRC5-dependent activation of the inflammasome. J Immunol 186(3):1333–1337. doi:10.4049/jimmunol.1003111

57. Biswas A, Meissner TB, Kawai T, Kobayashi KS (2012) Cutting edge: impaired MHC class I expression in mice deficient for Nlrc5/class I transactivator. J Immunol 189(2):516–520. doi:10.4049/jimmunol.1200064

58. Meissner TB, Li A, Biswas A, Lee KH, Liu YJ, Bayir E, Iliopoulos D, van den Elsen PJ, Kobayashi KS (2010) NLR family member NLRC5 is a transcriptional regulator of MHC class I genes. Proc Natl Acad Sci U S A 107(31):13794–13799. doi:10.1073/pnas.1008684107

59. Meissner TB, Li A, Kobayashi KS (2012) NLRC5: a newly discovered MHC class I transactivator (CITA). Microbes Infect 14(6):477–484. doi:10.1016/j.micinf.2011.12.007

60. Meissner TB, Li A, Liu YJ, Gagnon E, Kobayashi KS (2012) The nucleotide-binding domain of NLRC5 is critical for nuclear import and transactivation activity. Biochem Biophys Res Commun 418(4):786–791. doi:10.1016/j.bbrc.2012.01.104

61. Benko S, Magalhaes JG, Philpott DJ, Girardin SE (2010) NLRC5 limits the activation of inflammatory pathways. J Immunol 185(3):1681–1691. doi:10.4049/jimmunol.0903900

62. Cui J, Zhu L, Xia X, Wang HY, Legras X, Hong J, Ji J, Shen P, Zheng S, Chen ZJ, Wang RF (2010) NLRC5 negatively regulates the NF-kappaB and type I interferon signaling pathways. Cell 141(3):483–496. doi:10.1016/j.cell.2010.03.040

63. Tong Y, Cui J, Li Q, Zou J, Wang HY, Wang RF (2012) Enhanced TLR-induced NF-kappaB signaling and type I interferon responses in NLRC5 deficient mice. Cell Res 22(5):822–835. doi:10.1038/cr.2012.53

64. Neerincx A, Lautz K, Menning M, Kremmer E, Zigrino P, Hosel M, Buning H, Schwarzenbacher R, Kufer TA (2010) A role for the human nucleotide-binding domain, leucine-rich repeat-containing family member NLRC5 in antiviral responses. J Biol Chem 285(34):26223–26232. doi:10.1074/jbc.M110.109736

65. Ranjan P, Singh N, Kumar A, Neerincx A, Kremmer E, Cao W, Davis WG, Katz JM, Gangappa S, Lin R, Kufer TA, Sambhara S (2015) NLRC5 interacts with RIG-I to induce a robust antiviral response against influenza virus infection. Eur J Immunol 45:758–772. doi:10.1002/eji.201344412

66. Kumar H, Pandey S, Zou J, Kumagai Y, Takahashi K, Akira S, Kawai T (2011) NLRC5 deficiency does not influence cytokine induction by virus and bacteria infections. J Immunol 186(2):994–1000. doi:10.4049/jimmunol.1002094

Chapter 3

Assessment of Inflammasome Activation by Cytokine and Danger Signal Detection

Nicolas Riteau, Aurélie Gombault, and Isabelle Couillin

Abstract

The evaluation of the inflammasome activation usually addresses the presence of extracellular IL-1β and IL-18 or the secretion of danger signal proteins such as HMGB-1 through their quantification using an enzyme-linked immunosorbent assay (ELISA). The ELISA is a routine laboratory technique that uses antibodies and colorimetric changes to identify a substance of interest. ELISA uses a solid-phase enzyme immunoassay to detect the presence of a substance, usually an antigen, in a liquid or wet sample. Using 96 well plates, the ELISA technique enables to quantify the concentration of a single cytokine in multiple samples. However, a limitation of IL-1β and IL-18 ELISA is the absence of discrimination between active and non-active form of the proteins, parameter critical, for example, to distinguish the biologically relevant IL-1β from its poorly active form pro-IL-1β. This issue can be solved using western blots or immunoblots (IB), a common analytical procedure to detect the presence of different proteins in biological samples. Using denaturing conditions, IB allows the visualization of different sizes of the proteins of choice and is a commonly used technique in the inflammasome field to evaluate, for instance, the maturation of pro-IL-1β, pro-IL-18, and pro-caspase-1 into mature IL-1β, mature IL-18, and mature caspase-1, respectively. Moreover inflammasome activation may lead to the release of inflammasome particles outside the cell through caspase-1- or caspase-11-dependent cell death mechanism termed pyroptosis. In this case, NLR, ASC, and caspase-1 components are detectable outside the cell using IB analysis. ELISA and IB can be performed on cell culture supernatant or cell extract and on ex vivo samples from organ homogenates or biological fluids such as serum and plasma or bronchoalveolar lavages.

Key words ELISA, Immunoblot, IL-1β, IL-18, HMGB-1, ATP

1 Introduction

The inflammasomes are intracellular protein complexes that enable autocatalytic activation of inflammatory caspases (in particular caspase-1 and caspase-11) [1, 2]. They drive host and immune responses through the activation and secretion of pro-inflammatory cytokines and the release of intracellular derived components, which can be sensed by neighboring cells as danger signals or alarmins [2–5]. In addition, inflammasome activation can lead to pyroptosis, a pro-inflammatory type of cell death [6, 7].

Francesco Di Virgilio and Pablo Pelegrín (eds.), *NLR Proteins: Methods and Protocols*, Methods in Molecular Biology, vol. 1417, DOI 10.1007/978-1-4939-3566-6_3, © Springer Science+Business Media New York 2016

In general, inflammasome activation is tightly regulated in a two-step mechanism and involves the detection of microbial substances and/or danger signals leading to the activation of the inflammasome sensor molecule, such as Nod-like receptor 3 (NLRP3), followed by the recruitment and polymerization of the adaptor apoptosis-associated speck-like protein containing a caspase-recruitment domain (ASC). Consecutively, ASC recruits the pro-caspase-1 through its caspase activation and recruitment domain (CARD), and physical proximity of pro-caspase-1 induces its self-activation. Active caspase-1 in turn activates by proteolytic cleavage the members of the IL-1 cytokine family pro-IL-1β and pro-IL-18 into their biologically active forms IL-1β and IL-18 [8].

Moreover, the inflammasomes are also involved in the nonconventional secretion of leaderless proteins (in particular mature IL-1β and IL-18) as well as other proteins which are not direct caspase-1 substrates such as the high-mobility group box 1 (HMGB-1) and more than 20 other proteins. Some of the molecules released upon inflammasome activation are recognized as danger signals named also "danger-associated molecular patterns" (DAMPs), which are a set of host-derived molecules that signal cellular stress, damage, and among them, HMGB-1, uric acid, and ATP play major roles in eliciting inflammation and tissue repair during infection and under conditions of noninfectious (sterile) inflammation [9, 10]. Here, we describe two standard protocols to evaluate inflammasome activation through the detection of pro-inflammatory cytokine maturation and/or secretion or DAMP secretion evaluated by ELISA or IB.

2 Materials

Unless otherwise noted, prepare all solutions in distilled water at room temperature (*see* **Note 1**).

2.1 Reagents for Cell Culture and Stimulation

1. Differentiation cell culture media: Dulbecco's Modified Eagle Medium (DMEM) supplemented with 20 % horse serum and 30 % L929 cell-conditioned medium (*see* **Note 2**).

2. 100 ng/mL lipopolysaccharide (LPS) from *Escherichia coli* serotype 055:B5 in cell culture media.

3. OptiMEM cell culture media (*see* **Note 3**).

4. 20 μM nigericin in OptiMEM.

2.2 Reagents Necessary for ELISA

The following reagents illustrate the quantification of mouse IL-1β but can be adjusted to any protein of interest.

1. ELISA 96 well microplate.

2. Blocking buffer as supplied by the ELISA manufacturer.

3. Reagent diluent as supplied by the ELISA manufacturer.

4. Wash buffer: 0.05 % Tween 20 in phosphate-buffered saline solution (PBS).

5. Mouse IL-1β standard: prepare a seven-point standard curve of recombinant mouse IL-1β using twofold serial dilutions in reagent diluent.

6. Primary antibody: dilute mouse IL-1β capture antibody to a working concentration in PBS, without carrier protein as per recommendation of the manufacturer.

7. Secondary biotin labeled antibody: dilute mouse IL-1β detection antibody to the working concentration in reagent diluent as per recommendation of the manufacturer.

8. Streptavidin-HRP: dilute to HRP to the working concentration specified on the vial label using reagent diluent.

9. HRP substrate: 0.6 mM 2,2′-azino-bis(3-ethylbenzothiazoline-6-sulphonic acid (ABTS)) in 0.1 M citric acid buffer, pH 4.35 and 0.03 % H_2O_2 (*see* **Note 4**).

10. ABTS Stop solution: 50 % dimethylformamide (DMF) and 20 % sodium dodecyl sulfate (SDS) in water.

11. Plate reader able to read optical density at 450 nm, with wavelength correction at 540 nm or 570 nm.

2.3 Reagents Necessary for Western Blot or IB

1. 10 % ammonium persulfate (APS) solution in water (*see* **Note 5**).

2. Laemmli buffer: 50 mM Tris pH 6.8, 1 % SDS, 10 % glycerol, 0.125 mM β-mercaptoethanol, 0.005 % bromophenol blue.

3. 0.2 % Ponceau red with 3 % trichloroacetic acid (TCA) solution.

4. 10× running buffer: 250 mM Tris, 1.92 M glycine, 1 % SDS. To obtain 1× running buffer, dilute 10× running buffer in water.

5. 10× transfer buffer: 250 mM Tris, 1.92 M glycine. To obtain 1× transfer buffer, dilute 100 mL of 10× transfer buffer, 800 mL of water, and 100 mL of ethanol.

6. TBS 10×: 1.4 M NaCl, 0.2 M Tris pH 7.6.

7. Washing buffer (T-TBS): 0.1 % Tween 20 in TBS.

8. Blocking buffer: 10 % non-fat milk or bovine serum albumin in T-TBS.

9. 0.5 mM sodium azide.

10. Separating gel: reagents and volumes necessary to make one mini-gel are listed in Table 1 (*see* **Note 6**).

11. Stacking gel: reagents and volumes necessary to make one mini-gel stacking gel are listed in Table 2.

Table 1
Separating gel preparation. This table indicates the reagents and volumes required to prepare a separating gel, depending on the percentage of polyacrylamide desired. APS (ammonium persulfate) and TEMED have to be added at the end of the preparation

Acrylamide percentage	6 %	7 %	8 %	10 %	12 %	13 %	15 %
water (mL)	2.9	2.76	2.65	2.4	2.15	2	1.75
Acrylamide 40 % (mL)	0.75	0.875	1	1.25	1.5	1.63	1.9
Tris–HCl 1.5 M pH 8.8 (mL)	1.25	1.25	1.25	1.25	1.25	1.25	1.25
SDS 10 % (μL)	50	50	50	50	50	50	50
APS 10 % (μL)	50	50	50	50	50	50	50
TEMED (μL)	4	4	3	2.5	2	2	2

Table 2
Stacking gel preparation. This table indicates the reagents and volumes required to make a standard stacking acrylamide gel. APS (ammonium persulfate) and TEMED have to be added at the end of the preparation

Stacking	1 gel
Water	0.804 mL
Acrylamide 40 %	122 μL
Tris–HCl 0.5 M pH 6.8	313 μL
SDS 10 %	12.5 μL
APS 10 %	12.5 μL
TEMED	1.3 μL

3   Methods

3.1 Cell Stimulation and Sample Preparation

The following procedure illustrates the differentiation and activation of mouse bone marrow-derived macrophages (BMDMs) derived from C57BL/6 wild-type (B6) *Nlrp3* (*Nlrp3$^{-/-}$*) [11] or *Nlrp6* (*Nlrp6$^{-/-}$*) [12] deficient mice. However, other mouse genotypes can be used for this procedure. The Differentiation of bone marrow-derived macrophages is a process that has been explained in different publications [13]. In brief:

1. Isolate femoral bone from mice deficient for *Nlrp3*, *Nlrp6*, or B6 mice.

2. Flush femoral marrow with differentiation cell culture media.

3. Plate 10^6 of marrow cells/mL in differentiation cell culture media.

4. Incubate 3 days at 37 °C and 5 % CO_2.

5. Add 5 mL of fresh medium and culture the cells for further 4 days (*see* **Note 7**).

6. Plate the BMDMs in 96-well microculture plates at a density of 10^5 cells/well.

7. Stimulate with LPS during 3 h at 37 °C and 5 % CO_2.

8. Wash the cells with PBS and stimulate with or without nigericin (*see* **Note 8**) during 1 h at 37 °C and 5 % CO_2.

9. Collect cell supernatants and store at –20 °C for further analysis by ELISA or IB.

3.2 ELISA

The following procedure illustrates the quantification of mouse IL-1β but can be adjusted to any protein of interest.

1. Prepare all reagents as specified in Subheading 2.1 (*see* **Note 9**).

2. Coat the 96-well microplate with 100 μL per well of the diluted capture antibody solution. Seal the plate and incubate overnight at room temperature.

3. Aspirate each well and wash twice with wash buffer and wash at last with PBS for a total of three times (*see* **Note 10**). After the last wash, remove any remaining wash buffer by aspirating or by inverting the plate and blotting it against clean paper towels.

4. Block the plates by adding 300 μL of blocking buffer to each well.

5. Incubate at room temperature for a minimum of 1 h (*see* **Note 11**).

6. Repeat the aspiration/wash as in **step 3**; the plates are now ready for sample addition.

7. Add 100 μL per well of an appropriate dilution of the sample or standards in reagent diluent (*see* **Note 12**). Cover the plate with an adhesive strip and incubate for 2 h at room temperature or overnight at 4 °C.

8. Repeat the aspiration/wash as in **step 3**.

9. Add 100 μL of the detection antibody dilution to each well. Cover with a new adhesive strip and incubate 2 h at room temperature.

10. Repeat the aspiration/wash as in **step 3**.

11. Add 100 μL of the working dilution of streptavidin-HRP to each well. Cover the plate and incubate for 20 min at room temperature (*see* **Note 13**).

12. Repeat the aspiration/wash as in **step 3**.

13. Add 100 μL of HRP substrate solution to each well. Incubate for 20 min at room temperature (*see* **Notes 13** and **14**).

14. Add 50 μL of stop solution to each well. Gently tap the plate to ensure thorough mixing.

15. Determine the optical density of each well immediately, using a microplate reader set to 450 nm. If wavelength correction is available, set to 540 nm or 570 nm (*see* **Note 15**).

16. The optical density will be proportional to the initial amount of analyte present in the sample and can be quantified with the used of the standard curve developed from the same plate (*see* **Note 16**).

3.3 Chloroform/ Methanol Protein Precipitation for IB

There are many ways to concentrate proteins into smaller volumes in order to fit with the maximum loading capacity of the gels prior to electrophoresis and IB. Here, we describe two different methods to concentrate proteins: the chloroform/methanol precipitation (typically sample volume not to exceed 600 μL, Subheading 3.3) and the acetone precipitation (usually for larger sample volume, Subheading 3.4).

The chloroform/methanol precipitation is useful to concentrate cell culture supernatant as produced in Subheading 3.1. The following protocol is designed for a culture containing 10^6 adherent cells in a 12 well plate in final volume of 600 μL. Perform this procedure at room temperature and the centrifugation steps at 4 °C.

1. Stimulate macrophages in 600 μL of medium without serum (*see* **Note 17**).

2. Collect the cell culture supernatant in a microcentrifuge tube and spin down at $450 \times g$ for 5 min at 4 °C.

3. Transfer the supernatant to a new tube. Discard the other tube containing floating cells and debris.

4. Add 1 volume of sample supernatant with 1 volume of methanol and 0.25 volume of chloroform (e.g., for 600 μL of sample supernatant, add 600 μL methanol and 150 μL of chloroform).

5. Close the lid and mix vigorously for at least 30 s. At this step, the sample must become cloudy.

6. Spin down at $14,000 \times g$ for 10 min at 4 °C.

7. Remove the aqueous top layer without altering the interphase containing the proteins (thin white layer).

8. Add 100 μL methanol and mix gently until the proteins are settling to form a pellet (*see* **Note 18**).

9. Spin down the samples at $14,000 \times g$ for 10 min at 4 °C.

10. Carefully remove the supernatant by gentle aspiration (the pellet is not stable).

11. Let the microcentrifuge tube lid open and let dry upside down for a couple of minutes (*see* **Note 19**).

12. Add a desired volume (e.g., 30 µL) of reducing buffer 1×, e.g., laemmli buffer (*see* **Note 20**).

13. Use a pipette to homogenate the sample until the proteins are fully solubilized (*see* **Note 21**).

14. Store samples at –20 °C for further use or proceed to Subheading 3.5.

3.4 Acetone Protein Precipitation for IB

Acetone precipitation can be easily performed with higher sample volume. This can be useful, for example, when a low signal is expected and a large amount of sample needs to be concentrated.

1. Transfer cell culture supernatant or tissue homogenate obtained in Subheading 3.1 to a 15 mL canonical tube.

2. Add 3 volume of acetone to the samples (e.g., for 1 mL of sample, add 3 mL of acetone).

3. Close the tube and mix by inverting it.

4. Cool the tube for a minimum of two hours at –20 °C.

5. Spin down at $340 \times g$ for 10 min at 4 °C.

6. Carefully remove the supernatant.

7. Let the tube open and let dry upside down a couple of minutes (*see* **Note 19**).

8. Add a desired volume of reducing buffer of choice (e.g., 30 µL of laemmli buffer) (*see* **Note 20**).

9. Use a pipette to homogenate the sample until the proteins are fully solubilized (*see* **Note 21**).

10. Store samples at –20 °C for further use or proceed to subheading 3.5.

3.5 Western Blot or IB: All the incubation and washing steps must be performed on a rocking platform or a similar device.

1. Use precast gels or pour your own SDS-PAGE gel with appropriate percentage of acrylamide depending on the molecular weight of the proteins of interest (*see* **Note 6**).

2. Mix samples from Subheadings 3.1, 3.3 or 3.4 with loading buffer and boil at 95 °C for 5 min.

3. Spin down the tubes for a few seconds in a bench minicentrifuge.

4. Load equal amounts of protein into the wells of the SDS-PAGE gel along with molecular weight markers (*see* **Note 22**).

5. Run the gel, and the time and voltage need to be adjusted depending on the equipment used and the percentage of acrylamide of the gel (*see* **Note 23**).

6. Transfer the protein from the gel to the nitrocellulose membrane (*see* **Note 24**), which is equilibrated in transfer buffer for 15 min before transfer (*see* **Note 25**).

7. Block the membrane with T-TBS milk 10 % for 2 h at room temperature(*see* **Note 26**).

8. Incubate the membrane with the primary antibody in T-TBS milk 5 % overnight at 4 °C (*see* **Note 27**).

9. Wash the membrane with a large volume of T-TBS 1 % once quickly and then three times for a minimum of 5 min.

10. If necessary, incubate the membrane with a secondary antibody in T-TBS milk 5 % for 1 h at room temperature(*see* **Note 28**).

11. Wash the membrane with large volume of T-TBS 1 % once quickly and then three times for a minimum of 5 min (*see* **Note 29**).

12. If the antibody is coupled to HRP, add a chemiluminescence substrate to the membrane for 5 min at room temperature (e.g., ECL substrate).

13. Remove the excess of substrate.

14. Wrap the membrane in parafilm.

15. In a dark room, expose the membrane to an X-ray film and develop (*see* **Note 30**).

16. Example of membranes immunoblotted with a Rabbit polyclonal against HMGB-1 showed that the NLRP3 inflammasome activation by nigericin induces the secretion of the danger signal HMBG-1 by wild-type and *Nlrp6*⁻/⁻ BMDM but not by *Nlrp3*⁻/⁻ BMDM (*see* Fig. 1). These results confirm that HMGB-1 secretion is dependent of the NLRP3 inflammasome activation [14].

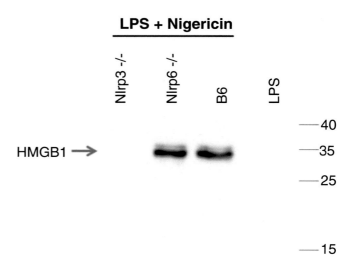

Fig. 1 HMGB-1 secretion by murine macrophages upon nigericin depends on NLRP3 but not on NLRP6. Western blot analysis of LPS-primed BMDM supernatants (SN) from wild-type, *Nlrp6*⁻/⁻ or *Nlrp3*⁻/⁻ mice confirmed that nigericin (20 μM) induces the secretion of HMGB-1 (30 kDa) in a NLRP3-dependant manner. Molecular weight markers are shown on the *right*

4 Notes

1. These buffers are commercially available.

2. Grow L929 cells in a 37 °C and 5 CO_2 humidified incubator using DMEM supplemented with 10 % of fetal calf serum for 7 days until confluence. Then filtrate the media through 0.2 µm filter and use it as a source of M-CSF to differentiate bone marrow-derived macrophages.

3. OptiMEM is an exclusive media from Life Technologies; other media without serum, i.e., DMEM, may be also suitable.

4. Tetramethylbenzidine (TMB) could be a substitute of ABTS.

5. APS solution is stable at 4 °C for 2 weeks (freezing is recommended for longer storage).

6. The percentage of acrylamide in the separating gel depends on the molecular weight of your protein of interest. For example, to detect IL-1β, knowing that the pro-form is 37 kDa and the mature form is 17 kDa, a 13 % acrylamide gel can be used (*see* Fig. 2).

7. After 7 days of culture, the cell preparation contained a homogenous population of >95 % of macrophages.

8. Nigericin is a K^+-ionophore and a well-known activator of the NLRP3 inflammasome.

9. Bring all reagents at room temperature before use. Allow all components to sit for a minimum of 15 min with gentle shaking

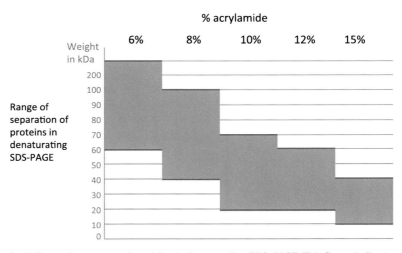

Fig. 2 Separation ranges of proteins in denaturing SDS-PAGE. This figure indicates the optimal percentage of acrylamide in the separating gel depending on the size of the target protein. For example, 6 % of acrylamide is better for 80–100 kDa protein whereas 15 % of acrylamide is for smaller protein such as 10–30 kDa (Adapted from *Roche Molecular Biochemicals*)

after initial reconstitution. Working dilutions should be prepared and used immediately.

10. Wash by filling each well with wash buffer (300 µL) using a squirt bottle, a manifold dispenser, or a plate washer. A complete removal of the liquid at each step is important for a good performance of the assay.

11. Blocking could last up to 16 h, but in that case, leave the blocking reaction at 4 °C.

12. From cell supernatants of BMDMs treated with LPS and nigericin (as described in Subheading 3.1), a 1:4 dilution is recommended to detect IL-1β.

13. Avoid placing the plate in direct light.

14. The HRP enzyme converts the substrate into another compound exhibiting different optical properties, and the optical density can be measured at specific wavelengths.

15. If wavelength correction is not available, subtract readings at 540 nm or 570 nm from the readings at 450 nm. This subtraction will correct the optical imperfections in the plate. Readings made directly at 450 nm without correction may give higher optical density values and be less accurate.

16. Take the averages of the duplicate values of the optical density for each concentration of recombinant protein and plot vs. the value of the concentration. Then fit a linear regression and use its equation to estimate the concentration of your test samples. Standard curve can be calculated using Microsoft Excel or GraphPad Prism software.

17. A suitable medium without serum is OptiMEM; alternatively, DMEM without serum can be also used. Excess of serum will result in a higher protein concentration in the precipitate and could generate non-specific background on the membrane.

18. It is recommended to mix gently to avoid proteins to stick to the sides of the microcentrifuge tube, which will make them harder to solubilize and may decrease the protein yield.

19. It is not necessary to completely dry the protein pellet, as it will be harder to resuspend.

20. It is possible to use another reducing buffer of choice or loading buffer without bromophenol blue which interferes with protein dosage, in order to measure protein concentration in sample and then adjust final volume between samples to load the same quantity of protein and add bromophenol blue.

21. This step is critical and may influence the quality of the western blotting.

22. The total protein concentration will vary depending on the sample and the abundance of the protein of interest, typically

20–30 µg of total protein from cell lysate, tissue homogenate, or concentrated cell culture supernatant.

23. Run at low voltage (about 90 V) until proteins enter in the stacking gel.

24. Polyvinylidene fluoride (PVDF) membranes could be used instead of nitrocellulose membranes.

25. After transfer, it is possible to stain the membrane with Ponceau red to assess total protein profile and transfer efficiency.

26. It is possible to block overnight at 4 °C.

27. Dilution ranges of primary antibodies usually range from 1:100 to 1:1000 in blocking buffer. After use, it is possible to store the antibody for further use at 4 °C by adding 0.5 mM sodium azide.

28. Dilution ranges of secondary antibodies usually range from 1:5000 to 1:10,000 in blocking buffer.

29. This washing step is critical to avoid non-specific signal.

30. Blots could also be developed using a cooled CCD camera, such as the Pxi machine (Ozyme) without film.

Acknowledgement

This work was supported by the following grants: «Fonds de Dotation pour la Recherche en Santé Respiratoire», the «Agence Nationale de la Recherche» and «Conseil Général du Loiret».

This research was supported in part by the Intramural Research Program of the NIH, NIAID.

References

1. Cerretti DP, Kozlosky CJ, Mosley B, Nelson N, Van Ness K, Greenstreet TA, March CJ, Kronheim SR, Druck T, Cannizzaro LA et al (1992) Molecular cloning of the interleukin-1 beta converting enzyme. Science 256(5053): 97–100

2. Kuida K, Lippke JA, Ku G, Harding MW, Livingston DJ, Su MS, Flavell RA (1995) Altered cytokine export and apoptosis in mice deficient in interleukin-1 beta converting enzyme. Science 267(5206):2000–2003

3. Nickel W, Rabouille C (2009) Mechanisms of regulated unconventional protein secretion. Nat Rev Mol Cell Biol 10(2):148–155

4. Rubartelli A, Cozzolino F, Talio M, Sitia R (1990) A novel secretory pathway for interleukin-1 beta, a protein lacking a signal sequence. EMBO J 9(5):1503–1510

5. Sutterwala FS, Ogura Y, Szczepanik M, Lara-Tejero M, Lichtenberger GS, Grant EP, Bertin

J, Coyle AJ, Galan JE, Askenase PW, Flavell RA (2006) Critical role for NALP3/CIAS1/Cryopyrin in innate and adaptive immunity through its regulation of caspase-1. Immunity 24(3):317–327

6. Chen Y, Smith MR, Thirumalai K, Zychlinsky A (1996) A bacterial invasin induces macrophage apoptosis by binding directly to ICE. EMBO J 15(15):3853–3860

7. Lamkanfi M, Dixit VM (2010) Manipulation of host cell death pathways during microbial infections. Cell Host Microbe 8(1):44–54

8. Martinon F, Burns K, Tschopp J (2002) The inflammasome: a molecular platform triggering activation of inflammatory caspases and processing of proIL-beta. Mol Cell 10(2):417–426, pii:S1097276502005993

9. Lamkanfi M (2011) Emerging inflammasome effector mechanisms. Nat Rev Immunol 11(3): 213–220

10. Lamkanfi M, Dixit VM (2012) Inflammasomes and their roles in health and disease. Annu Rev Cell Dev Biol 28:137–161

11. Martinon F, Petrilli V, Mayor A, Tardivel A, Tschopp J (2006) Gout-associated uric acid crystals activate the NALP3 inflammasome. Nature 440(7081):237–241

12. Chen GY, Liu M, Wang F, Bertin J, Nunez G (2011) A functional role for Nlrp6 in intestinal inflammation and tumorigenesis. J Immunol 186(12):7187–7194

13. Jiang D, Liang J, Fan J, Yu S, Chen S, Luo Y, Prestwich GD, Mascarenhas MM, Garg HG, Quinn DA, Homer RJ, Goldstein DR, Bucala R, Lee PJ, Medzhitov R, Noble PW (2005) Regulation of lung injury and repair by Toll-like receptors and hyaluronan. Nat Med 11(11):1173–1179

14. Dupont N, Jiang S, Pilli M, Ornatowski W, Bhattacharya D, Deretic V (2011) Autophagy-based unconventional secretory pathway for extracellular delivery of IL-1beta. EMBO J 30(23):4701–4711

Chapter 4

Investigating IL-1β Secretion Using Real-Time Single-Cell Imaging

Catherine Diamond, James Bagnall, David G. Spiller, Michael R. White, Alessandra Mortellaro, Pawel Paszek, and David Brough

Abstract

The pro-inflammatory cytokine interleukin (IL)-1β is an important mediator of the inflammatory response. In order to perform its role in the inflammatory cascade, IL-1β must be secreted from the cell, yet it lacks a signal peptide that is required for conventional secretion, and the exact mechanism of release remains undefined. Conventional biochemical methods have limited the investigation into the processes involved in IL-1β secretion to population dynamics, yet heterogeneity between cells has been observed at a single-cell level. Here, greater sensitivity is achieved with the use of a newly developed vector that codes for a fluorescently labelled version of IL-1β. Combining this with real-time single-cell confocal microscopy using the methods described here, we have developed an effective protocol for investigating the mechanisms of IL-1β secretion and the testing of the hypothesis that IL-1β secretion requires membrane permeabilisation.

Key words Confocal microscopy, IL-1β secretion, Lentiviral transduction, IL-1β Venus, Real-time single-cell imaging

1 Introduction

Cytokine secretion through nonconventional pathways has been a subject of great interest in recent years. Interleukin-1β (IL-1β) secretion has been a particular research focus as it is a major regulator of the inflammatory response. IL-1β is produced mainly by cells of macrophage lineage as a precursor, pro-IL-1β. Pro-IL-1β is expressed in response to pathogen-associated molecular patterns (PAMPs) or damage-associated molecular patterns (DAMPs) that stimulate pattern recognition receptors (PRRs) on macrophages. PAMPs are molecules expressed by pathogens, such as bacterial lipopolysaccharide (LPS) of Gram-negative bacteria, and DAMPs are typically endogenous molecules released by cellular necrosis or endogenous molecules modified during disease. Pro-IL-1β is inactive and a further stimulation of the activated macrophage by

Francesco Di Virgilio and Pablo Pelegrín (eds.), *NLR Proteins: Methods and Protocols*, Methods in Molecular Biology, vol. 1417, DOI 10.1007/978-1-4939-3566-6_4, © Springer Science+Business Media New York 2016

an additional PAMP or DAMP is required to activate cytosolic PRRs, often of the NLR family, to form large multi-protein complexes called inflammasomes. Inflammasomes activate the protease caspase-1 that drives processing of pro-IL-1β and release of the mature cytokine [1]. The best characterised inflammasome is formed by the PRR NLRP3 (NACHT, LRR and PYD domains-containing protein 3).

IL-1β is known to have multiple effects on the immune response, yet the mechanisms involved in its secretion remain to be resolved. It is almost 30 years since the discoveries that IL-1β lacks a signal peptide [2] and does not traffic through the ER and Golgi [3], and we still do not understand how IL-1β is secreted. There are however a number of possible secretory routes that have been described, as we have reviewed recently [1]. These mechanisms can be classed as either rescue and redirect, where IL-1β targeted to the lysosome for degradation is diverted to the extracellular space via exocytosis; protected release, where IL-1β is released in discrete packages which can be either plasma membrane-derived microvesicles or exosomes released from multivesicular bodies; or terminal release, where IL-1β is passively released through pores in the plasma membrane, which is related to pyroptotic cell death. The vast majority of studies leading to these observations have relied on studying population dynamics using antibody-based assays. However, this may not fully represent how single cells react to stimuli. Recent studies have demonstrated the heterogeneity of cytokine secretion within a population [4], and this has confirmed the need for single-cell analysis that allows spatiotemporal determination of the mechanisms involved in cytokine secretion. Emerging technology is allowing more thorough investigation of protein secretion including real-time single-cell imaging using confocal microscopy, which we describe below.

Here we used real-time single-cell imaging using confocal microscopy to establish a protocol that allows visualisation of the nonconventional secretion pathway utilised by IL-1β. Using the methods described below, we were able to provide evidence that membrane permeability occurs concomitantly with the release of IL-1β, suggesting that loss of membrane integrity is a requirement for the secretion of the cytokine under these conditions. An example of the data is shown in Fig. 1. These data are in agreement with the results from two recent studies, where variations of single-cell live-cell imaging methods were used to interrogate the cytokine secretion pathway. Shirasaki et al. (2014) used total internal reflection fluorescence microscopy (TIFRM) to monitor a sandwich immunoassay performed on individual cells, to provide evidence suggesting that IL-1β secretion relies on the loss of membrane integrity [5]. This group also developed SCAT1 that enabled the detection of caspase activation using fluorescence microscopy. The fluorescence resonance energy transfer (FRET) between two

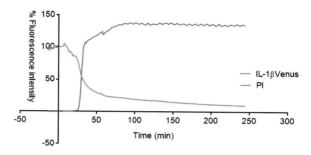

Fig. 1 Time course of the average intensity of fluorescence of IL-1β-Venus (*green*) and PI (*red*) (normalised to baseline), when BMDM IL-1β-Venus cells were treated with ATP (5 mM) (time = 0) with prior stimulation by 1 μg/ml LPS 4 h

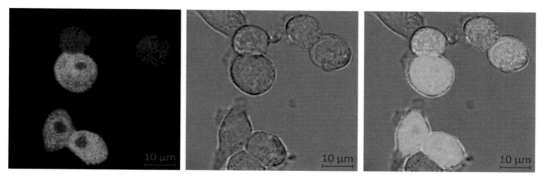

Fig. 2 Immortalised mouse BMDMs expressing IL-1β-Venus. The bright-field image, the corresponding fluorescence image and the merged channel (both bright-field and fluorescence) images are shown. IL-1β-Venus is distributed evenly across the cytoplasm. Scale bar represents 10 μm

fluorescent proteins is disturbed upon activation of caspase-1, as the sequence of the linker between them contains caspase-1 cleavage sites. This technique was utilised to show that IL-1β is released in a local burst as a result of a caspase-1 activation being an all-or-none response at the single-cell level [6].

To study the mechanism of IL-1β release, we transformed immortalised mouse bone marrow-derived macrophages (BMDMs) to express pro-IL-1β with a C-terminal fluorescent Venus tag, using a lentiviral approach (Fig. 2). We devised a protocol based on this cell line to visualise the real-time secretion of IL-1β from single cells, as seen in Fig. 1. With confocal live-cell microscopy, we were able to visually monitor how priming of the BMDMs with LPS and then the use of adenosine triphosphate (ATP) as a second stimulus induced the secretion of IL-1β. ATP is a commonly used stimulus in research into IL-1β secretion as it activates the P2X7 receptor on LPS-primed macrophages, causing an efflux of K^+ which is required for the assembly of the inflammasome and the activation of caspase-1. This subsequently leads to the release of IL-1β [7–9].

To simultaneously monitor membrane integrity, we included the cell-impermeant nucleic acid stain propidium iodide (PI) in the cell culture medium. Using the methods described in detail below, we highlight how real-time single-cell imaging using confocal microscopy can be a powerful tool for investigating the processes involved in nonconventional protein secretion.

2   Materials

The cell type used for the visualisation of IL-1β secretion was an immortalised bone marrow-derived macrophage (BMDM) cell line transduced to stably express IL-1β tagged with the fluorescent protein Venus (IL-1βVenus) using methods described in Subheadings 3.1–3.4.

2.1   Cloning

1. Zeocin agarose plates. Add to standard agarose plates 50 μg/ml of zeocin.

2. Gateway® pDONR/zeo vector (this vector is exclusive of Life Technologies).

3. pLNT-UbC-#-Venus vector, the custom third-generation lentiviral vector used here, which is currently only available in Paszek's lab, Manchester, UK (# denotes the insertion/recombination point).

4. 7.5 mM polyethylenimine (PEI) 'Max' (Polysciences, Inc.) solution: prepare it by mixing 63 μl PEI with 937 μl 150 mM NaCl solution.

2.2   Maintenance of Cells

1. Cell media: Dulbecco's modified Eagle's medium (DMEM) supplemented with 10 % heat-inactivated foetal bovine serum (FBS), 1 % (2 mM) L-glutamine Q and 1 % penicillin-streptomycin.

2. Cells are maintained at 37 °C in a humidified incubator (5 % CO_2, 95 % air).

3. 35 mm glass-bottomed microscopy plates (Greiner Bio-One).

4. 75 cm² tissue culture flask (T-75).

5. Sterile phosphate-buffered saline solution (PBS).

6. Sterile cell scrapers.

7. 15 ml conical falcon tubes.

8. Centrifuge.

2.3   Stimulation of IL-1β Release

1. 1 mg/ml of lipopolysaccharide (LPS) stock solution in PBS. Use a 1:1000 dilution to achieve a final concentration of 1 μg/ml (as described in Subheading 3.4).

2. 500 mM ATP stock solution using sterile distilled water, adjust pH to 7.0 using 5 M NaOH. Store stock solution at −20 °C until use. Use a 1:100 dilution to give a final concentration of 5 mM.

3. 1 mg/ml propidium iodide (PI) stock solution in water. Use a 1:200 dilution for a final concentration of 5 μg/ml. 0.75 μl was added to a 0.5 ml tube, then 7–10 μl of cell medium was taken from the cell chamber and mixed with the PI before returning both the PI and medium to the well. This caused minimum disturbance to the cells and also allowed quicker diffusion of PI through the media (Subheading 3.6).

4. 10 % triton x-100 intermediate solution to dilute 1:100 and obtain a 0.1 % triton x-100.

2.4 Visualisation of the Cells

A range of microscopes and objectives are appropriate for visualisation of BMDMs expressing Venus-tagged pro-IL-1β. Here, Zeiss 710 and 780 confocal microscopes were used, which employed detector arrays to collect appropriate emission signals following excitation of the Venus fluorophore with an argon laser at 514 nm and 560 nm for PI. Image capture was performed using the Zeiss software, 'Zen 2010b SP1'. Fluar 40× NA 1.3 (oil immersion) and plan-apochromat 63× NA 1.4 objectives were used for all imaging.

2.5 Image Analysis

Cell Tracker (version 0.6) was used to quantify time-lapse confocal images [10, 11].

3 Methods

Carry out all procedures at room temperature.

3.1 Cloning the IL-1β Lentiviral Expression Vector

The amplification of murine IL-1β (accession number NM_008361) can be carried out using a variety of commercially available kits which use proofreading polymerases. Here, the IL-1β sequence (mouse) was amplified from a plasmid using polymerase chain reaction (PCR), and primers flanked by gateway recombination sequences (*see* Table 1). Once amplified, the PCR reaction can be run on an agarose gel and appropriate size band excised (size of IL-1β).

Table 1
Forward and reverse IL-1β primers: small letters = recombination gateway sequences and capitals equate to IL-1β sequences

Forward IL-1β	gggg aca agt ttg tac aaa aaa gca ggc ttc acc ATGGCAACTGTTCCTGAACTC
Reverse IL-1β	gggg ac cac ttt gta caa gaa agc tgg gtc GGAAGACACGGATTCCATGGT

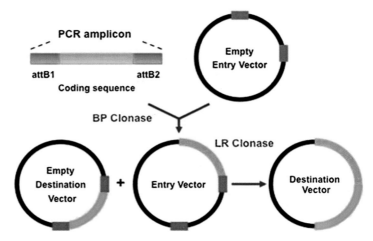

Fig. 3 Schematic representation of the Gateway® Invitrogen homologous recombination cloning strategy, which was employed to generate plasmids encoding fluorescent-tagged proteins. Orange indicates coding sequence of interest (for this method, IL-1β) and green denotes fluorescent protein coding sequence (here Venus plasmid). The recombination sites are shown as blue and red [13]

After excision the amplicon DNA can be purified from the agarose using an agarose gel extraction kit. This yields the purified amplicon.

The amplified IL-1β gene was then transferred to the 'pLNT' destination vectors using the standard Gateway cloning protocols [12], and therefore the steps are only described briefly here (*see* also Fig. 3 for graphical overview of cloning process).

1. BP reaction: Recombination to introduce the sequence into the Gateway® pDONR/zeo entry vector using 150 ng IL-1β amplicon + 150 ng entry vector made to a volume of 10 μl ddH$_2$O.

2. Incubate reaction at 25 °C for 3–24 h.

3. Transform 5 μl of this reaction in ccdB-sensitive competent cells, plating on zeocin plates (50 μg/ml).

4. Plasmid prep clones using standard kits and sequence the insert. This generates the terminal entry vector.

5. LR reaction: Recombination to transfer the gene to a pLNT-#-Venus expression vector, using 150 ng terminal entry vector + 150 ng destination vector (pLNT-#-venus).

6. Repeat **steps 2–4** as above.

The resultant vector is termed 'pLNT-IL-1β-Venus', which is a third-generation lentiviral transfer vector that constitutively expresses, from the ubiquitin-ligase C promoter, a Venus fluorescent protein fused to the C-terminus of the murine IL-1β protein. All of the cloning steps were validated by sequencing the recombination insertion site.

3.2 Lentivirus Production

1. Seed a 15 cm dish with 1.25×10^7 HEK293T cells using 20 ml media and incubate overnight at 37 °C.

2. Prepare polyethylenimine (PEI) 'Max' solution and store it at room temperature for 10 min.

3. Prepare 10.5 μg total DNA, made up by the packaging vectors pMDLg-RRE, pCMV-VSVG, pRSV-REV and the pLNT-IL-1β-Venus' transfer vector at a ratio of 2:1:2:4, making the total volume up to 1000 μl using NaCl. Store at room temperature for 10 min.

4. Combine the 1 ml of PEI solution with 1 ml of the DNA mix, to make the transfection mix. Wait for 10 min.

5. Add the PEI: DNA mix to the cells in a dropwise manner creating a 2 ml total volume. Swirl gently and leave for 10 min. This is the transfection mixture.

6. After 6 h, remove the transfection mixture and replace with 20 ml fresh media.

7. After 48 h, collect the cell media.

8. Concentrate the virus by ultracentrifugation of the viral media at $90,000 \times g$ for 90 min. Remove the supernatant and resuspend the virus particles in small volumes (100–200 μl) of 1×PBS and leave overnight.

3.3 Cell Transduction

1. Apply 100 μl of the virus to a culture of 1.5×10^4 immortalised BMDM cells in 2 ml of media for 48 h (the virus can be frozen at −80 °C for future use, but this can also reduce the titer).

2. After this period, replace the virally loaded media with fresh media and propagate the transduced culture for several passages.

3. Transduction efficiency can be determined by confocal microscopy.

The transduced cell culture can be frozen into stocks for use in all future experiments (media + 10 % DMSO, freeze cells at a rate of −1 °C per min).

3.4 Seeding the Microscopy Plates

Seed the cells onto 35 mm glass-bottomed microscopy plates the night before confocal microscopy imaging is due to be performed.

1. Aspirate the media from the 75 cm² tissue culture flask containing the cells grown to confluency.

2. Add 3 ml PBS to the flask, incubate at 37 °C for 2 min and scrape sides of flask to remove cells, with a sterile cell scraper.

3. Wash cells with PBS using a stripette and then transfer to a 15 ml falcon tube.

4. Use a cell counter at this stage to count the cells.

5. Centrifuge the cells for 5 min at $380 \times g$.

6. Aspirate PBS and gently resuspend the cells in 1 ml of prepared cell media (*see* Subheading 2.2).

7. Add the calculated amount of media needed, based on the density the cells are to be seeded at, and the number of wells (*see* **Note 1**). Here we used 500,000 cells/ml.

8. Add appropriate amount of media/cell mix to a glass-bottomed petri dish (depending on size of well). For the 35 mm dish used here, 3 ml total media was used (*see* **Note 2**).

9. Incubate overnight as stated above (*see* Subheading 2.2).

10. Add 1 µg/ml of LPS to each well needed for imaging and incubate for a further 4 h (stagger LPS stimulation if several wells are going to be used for consecutive imaging experiments).

3.5 Zeiss LSM 710/780 Setup

1. Switch on 514 and 560 argon lasers (these may be same laser depending on the system) and ensure environmental conditions are set to 37 °C, 5%CO_2. Load up LSM software.

2. Apply oil to objective if necessary.

3. Stabilise cell dish in plate holder and make sure it is completely level before securing. Check that the plate holder is flat by gently pressing down on one side and checking that the oil does not move.

4. In the 'Ocular' tab (Fig. 4), select 'Online' and 'Open' to use the bright field to locate cells (*see* **Note 3**). Adjust the focus manually until the cell membrane appears in focus (*see* **Note 4**).

5. When cells are in focus, switch to the 'Acquisition' tab (Fig. 5).

6. Ensure 'channel mode' is selected and add two tracks in 'Imaging setup', for EYFP (Venus) and PI (click '+' to add each track) (Fig. 6).

Fig. 4 'Ocular' icon

Fig. 5 'Acquisition' tab

Fig. 6 'Imaging setup'. Add the appropriate tracks for the fluorophores/fluorescent dyes. Here EYFP (for Venus) and PI. T-PMT represents the bright-field track

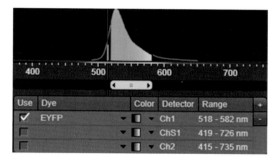

Fig. 7 Emission detection range. In the 'Light Path' window, set the correct wavelengths necessary for the fluorophores/fluorescent dyes

Fig. 8 Bright-field setup. Check the 'T-PMT' box to add bright-field imaging

7. Set appropriate wavelength detection under the 'Light path' setup by changing the emission detection range to the correct values (*see* **Note 5**) (Fig. 7).

8. Ensure the correct light path is set up to allow the correct wavelengths of light through to the detectors. This can also be done automatically using the 'Smart setup' underneath the 'Acquisition' tab, choosing the dyes you require. Check the T-PMT (transmitted light photomultiplier tube) box in the Venus channel to record the bright-field images (Fig. 8).

9. By clicking any of the four icons with the camera picture, 'Auto exposure', 'Live', 'Continuous', or 'Snap' (Fig. 9), the cells should be visible (*see* **Note 6**); however, more adjustments may need to be made in order to either make the cells focussed or to maximise the amount of information gained and the clarity of the cells.

Fig. 9 Camera scanning functions

10. Under the 'Channels' tab, adjust the 488 nm laser power percentage to appropriate levels that ensure adequate fluorescence levels to be able to distinguish the IL-1β-Venus fluorescing cells. However, this should be the minimum percentage possible to limit photobleaching and avoid saturation. Here, 6 % laser power was used.

11. Also under the 'Channels' tab, master detector gain level can be adjusted to optimise a balance with minimum laser power possible with no background noise detection. This value was between 700 and 900 for these experiments and can be checked using the range indicator palette (*see* **Note 7**).

12. The pinhole diameter was set such that fluorescence from the entire depth of the cells was collected; typically 3.4 airy units were used with a fluar 40× 1.3NA oil objective.

13. Repeat **steps 9–11** for each channel created. Here this will be only one repeat for PI; however, gain level should also be adjusted for bright field. The pinhole should be the same for all tracks/lasers used.

14. Once the fluorescent levels are optimised, adjust the focus manually, using the fluorescence as a guide. Usually, the maximum fluorescent levels are best, although again avoid saturation. Also, you can crop the background to gain a 'close up' of a region of interest.

15. Resolution can be changed under the 'Acquisition mode' tab. Depending on the speed at which the frames need to be imaged, a balance can be determined with the resolution; i.e. a higher resolution will give greater detail, but take longer to image. The averaging and speed at which the frames are taken also add to this careful balance, so be sure these are satisfactory before beginning the main experiment.

3.6 Time Series Imaging of IL-1β Release

1. Ensure the 'Time series' box is checked underneath the camera functions (Fig. 10). The number of frames needed and time interval between the frames should be entered underneath the 'Time series' tab.

2. Once the settings are correct, the experiment can begin. The cells should have had between 15 and 30 min to adjust to any perturbations in temperature or conditions.

3. If using PI, it should be added at this point (*see* Subheading 2.3).

Fig. 10 Imaging options

Fig. 11 'Autosave' window

Fig. 12 'Start experiment' button. This button becomes 'Stop experiment' once imaging has started

4. If using multiple positions/regions of interest (ROI), click 'Positions' in addition to 'Time series' underneath the camera functions. Mark the positions by manually finding a position of interest by manoeuvring the plate over the objective using the manual controller. At each ROI, click 'Mark/Add position'. This should mark the current position of x, y and z axis. To perform a check of the focus of each position after its addition, click on each position listed and press 'move to'. Each position should still be in focus (*see* **Note 8**).

5. It is also advisable to turn on 'Autosave' in the 'Autosave' tab (Fig. 11) and choose the appropriate file.

6. Ensure the room lights are switched off and press 'Start experiment' (Fig. 12).

7. While running 15 min of control imaging (no additional treatments at this point), ensure that no or very little PI can be seen (a control experiment can be conducted beforehand, only using triton-x100 detergent in PI-treated cells to allow adjustment to optimum PI levels).

8. After 5 min, click 'Pause' in the 'Time series' tab, and as quickly as possible, lift the chamber lids without moving the wells/plate; add 5 mM ATP to the well (final concentration).

Place the pipette tip on the edge of the well so that the ATP drips down the side and into the well. Again, this is to ensure minimum disturbance for the focus of each position.

9. Immediately press 'Resume' where the 'Pause' button was.

10. Check the focus of each position has been maintained (*see* **Note 8**) as the frames are imaged.

11. After 60 min, press 'Pause' again and add 0.1 % triton x-100 to each well used. This causes total cell lysis, acting as a control to see maximum permeability of the cells and ensures entry of PI into the cell.

12. After approximately 5 min (or until maximum PI fluorescence is reached), press 'Stop experiment' underneath the camera icons.

3.7 Cell-Tracking Analysis

1. There are many software packages that can be used to track cells. Here, image analysis was conducted using 'Cell Tracker' software. This requires prior installation of MATLAB.

2. Open the software and click the file icon. Browse for the file you require and press 'Load'.

3. The software can track cells automatically. However, often the software will not detect the correct cell or its border, and therefore the manual tracking system is described below. This can be used to modify the automatic tracking or to do without automatic tracking.

4. Using the spine tool, create a border around the cell to be tracked in the first frame. This can be edited using the spine editor tool (*see* **Note 9**).

5. If the cell is likely to be in the same place or a similar place in the next frame, right click the cell and choose 'copy' and choose the number of frames you want to copy the border to. Move to the next frame and modify the border if necessary. Repeat throughout all the image frames ensuring the cell border is measured, including any blebs that may occur in the membrane as the cell undergoes cell death.

6. Export the data as mean fluorescence intensity.

4 Notes

1. Confluency of cells needed for imaging: Depending on the cell type being used, the confluency needed for the clearest imaging is going to differ. Here BMDMs were used. This cell type requires a certain level of contact to maintain their phenotype. However, if cells are too confluent, they begin to cluster and form 'clumps' on different focal planes, making it very difficult

to track the cells. Here, the cells were plated at 500,000 cells/ ml and incubated overnight. Ideally, the ROI should be chosen to include several cells (here between 10 and 50) both isolated cells and those in groups of cells to decrease the chances of misleading images. During cell analysis, the whole cell should be clearly visible in every frame.

2. If running overnight experiments or experiments over 4 h, add a larger amount of media to compensate for any evaporation that may occur.

3. If the light does not seem bright across the whole field, use the condenser adjustments to move the condenser until you can see the edge of the field (dark arc) and then keep closing the condenser to get the field of view into the middle (the whole image is bright).

4. When trying to identify a ROI or focus the microscope on the ROI, start at the bottom corner of the well and work inwards. The plastic walls of the dish/plate distort the image, so slowly move towards the centre of the well and focus on the cells.

5. Use of detectors, PI and GFP emission wavelengths do not overlap, so the same detector can be used to detect them without getting overspill of fluorescence. This makes the overall imaging quicker. Here the wavelengths detected were set at: EYFP 505–550 nm, PI 580–680 nm.

6. Camera scanning functions: 'Autoexposure'—automatically adjusts the offset and detector gain for current laser power being used. 'Live'—live is continuous imaging at a reduced resolution. This is useful for finding the position of cells, but the resolution is too low to use to get a true focus on the cells. 'Snap'—one frame at the set resolution and speed. 'Continuous'—continuous imaging at the set quality and speed; however, it is not recording the images.

7. Range indicator: Click on the range indicator button below the image window so that the image appears in grey scale. If the pixels are too saturated, they appear red and their intensity cannot be quantified. Pixels that are at the minimum brightness (at 0) will appear blue. Adjust the laser power and gain until the image is hardly saturated.

8. Troubleshooting focus issues: If using multiple positions/ ROIs during the time series, issues with focus may be encountered. Check the plate/dish of cells is completely flat and stable as this may only be noticeable when the objective moves to each position.

9. Marking cell borders: it is advisable not to make too many spine points around the cell to create your border as this will cause modifications to become difficult. Between 5 and 9 points per cell depending on cell size is usually appropriate.

Acknowledgements

The authors are grateful to Prof Claire Bryant from the University of Cambridge for providing the immortalised BMDMs. P.P. holds a BBSRC David Phillips Research Fellowship (BB/I017976/1). This work was also supported by M.R.C. (MR/K015885/1) and BBSRC (BB/K003097/1). The work leading to these results has received funding from the European Union Seventh Framework Programme (FP7/2012-2017) under grant agreement n° 305564.

References

1. Lopez-Castejon G, Brough D (2011) Understanding the mechanism of IL-1beta secretion. Cytokine Growth Factor Rev 22: 189–195

2. Auron PE, Webb AC, Rosenwasser LJ, Mucci SF, Rich A, Wolff SM, Dinarello CA (1984) Nucleotide sequence of human monocyte interleukin 1 precursor cDNA. Proc Natl Acad Sci U S A 81:7907–7911

3. Rubartelli A, Cozzolino F, Talio M, Sitia R (1990) A novel secretory pathway for interleukin-1 beta, a protein lacking a signal sequence. EMBO J 9:1503–1510

4. Ma C, Fan R, Ahmad H, Shi Q, Comin-Anduix B, Chodon T, Koya RC, Liu CC, Kwong GA, Radu CG, Ribas A, Heath JR (2011) A clinical microchip for evaluation of single immune cells reveals high functional heterogeneity in phenotypically similar T cells. Nat Med 17:738–743

5. Shirasaki Y, Yamagishi M, Suzuki N, Izawa K, Nakahara A, Mizuno J, Shoji S, Heike T, Harada Y, Nishikomori R, Ohara O (2014) Real-time single-cell imaging of protein secretion. Sci Rep 4:4736

6. Liu T, Yamaguchi Y, Shirasaki Y, Shikada K, Yamagishi M, Hoshino K, Kaisho T, Takemoto K, Suzuki T, Kuranaga E, Ohara O, Miura M (2014) Single-cell imaging of caspase-1 dynamics reveals an all-or-none inflammasome signaling response. Cell Rep 8:974–982

7. Perregaux D, Gabel CA (1994) Interleukin-1 beta maturation and release in response to ATP and nigericin. Evidence that potassium

depletion mediated by these agents is a necessary and common feature of their activity. J Biol Chem 269:15195–15203

8. Ferrari D, Chiozzi P, Falzoni S, Dal Susino M, Melchiorri L, Baricordi OR, Di Virgilio F (1997) Extracellular ATP triggers IL-1 beta release by activating the purinergic P2Z receptor of human macrophages. J Immunol 159:1451–1458

9. Mariathasan S, Weiss DS, Newton K, McBride J, O'Rourke K, Roose-Girma M, Lee WP, Weinrauch Y, Monack DM, Dixit VM (2006) Cryopyrin activates the inflammasome in response to toxins and ATP. Nature 440: 228–232

10. Shen H, Nelson G, Nelson DE, Kennedy S, Spiller DG, Griffiths T, Paton N, Oliver SG, White MR, Kell DB (2006) Automated tracking of gene expression in individual cells and cell compartments. J R Soc Interface 3: 787–794

11. Du CJ, Marcello M, Spiller DG, White MR, Bretschneider T (2010) Interactive segmentation of clustered cells via geodesic commute distance and constrained density weighted Nystrom method. Cytometry A 77: 1137–1147

12. Katzen F (2007) Gateway((R)) recombinational cloning: a biological operating system. Expert Opin Drug Discov 2:571–589

13. Bagnall J (2011) Single-cell imaging and mathematical modelling of the hypoxia-inducible factor signalling network. PhD thesis, University of Liverpool

Chapter 5

Measuring IL-1β Processing by Bioluminescence Sensors I: Using a Bioluminescence Resonance Energy Transfer Biosensor

Vincent Compan and Pablo Pelegrín

Abstract

IL-1β processing is one of the hallmarks of inflammasome activation and drives the initiation of the inflammatory response. For decades, Western blot or ELISA have been extensively used to study this inflammatory event. Here, we describe the use of a bioluminescence resonance energy transfer (BRET) biosensor to monitor IL-1β processing in real time and in living macrophages either using a plate reader or a microscope.

Key words BRET, IL-1β, Sensor, Bioluminescence, Macrophage

1 Introduction

Inflammation is a physiological process in response to tissue injury or pathogen infection [1]. IL-1β, a proinflammatory cytokine, constitutes one of the key components involved in the biological cascade leading to inflammation. Different cell types, such as macrophages and monocytes, synthesize IL-1β as an inactive precursor [2]. Pro-IL-1β is processed to its active form by the protease caspase-1 and then is secreted to the extracellular space where it initiates the inflammatory response after binding to the IL-1 receptor located on neighboring cells [2, 3]. Caspase-1 activation is controlled by the assembly of Nod-like receptors (NLR) into multiprotein complexes termed inflammasomes [3]. A role for IL-1β has been reported in different pathologies including cancer, type 2 diabetes, gout, or Alzheimer's disease. Due to its predominant role in various pathophysiological processes, IL-1β processing has been studied employing antibodies against IL-1β in techniques such as Western blot or ELISA. These approaches have been extensively used for decades to detect IL-1β processing, and they are commonly used as an indirect method to detect NLR assembly into

Francesco Di Virgilio and Pablo Pelegrín (eds.), *NLR Proteins: Methods and Protocols*, Methods in Molecular Biology, vol. 1417, DOI 10.1007/978-1-4939-3566-6_5, © Springer Science+Business Media New York 2016

active inflammasomes in response to stimulation. However, these tools are limited in their temporal resolution and are not compatible for in situ IL-1β detection. We recently engineered a biosensor based on bioluminescence resonance energy transfer (BRET) that allows monitoring IL-1β processing in real time and in living cells, including macrophages. The precursor pro-IL-1β has been fused at its extremes to the donor (RLuc8) and the acceptor (Venus) of BRET, leading to an energy transfer between the two BRET partners (Fig. 1). We found that this biosensor has similar properties than its endogenous homolog pro-IL-1β and that it is cleaved by caspase-1 which leads to a decrease in BRET signal [4]. Using this biosensor, we were able to analyze IL-1β processing in real time by monitoring BRET variation in different macrophage cell lines. This tool is also compatible with microscopy to visualize such process on single cell as primary macrophage. This makes this biosensor an interesting tool for simple detection of IL-1β processing in situ.

2 Materials

Prepare all solutions using ultrapure water and analytical grade reagents. Prepare and store all reagents at room temperature (unless indicated otherwise). Diligently follow all waste disposal regulations when disposing waste materials. We do not add sodium azide to the reagents.

2.1 Expression of the Bioluminescence Sensor in Macrophage Cell Lines

1. Transfection reagent to transfect macrophage cell lines as J774A.1 or immortalized bone marrow derived macrophages. We use TransIT-Jurkat (Mirus) since it gives us good transfection on macrophages, but an equivalent reagent can be also used.

2. Complete cell media appropriate to the cell line in culture. For example, Dulbecco's Modified Eagle's Medium (DMEM) with 10 % of heat-inactivated fetal calf serum (FCS) for J774A.1.

3. Opti-MEM® culture media (Life Technologies) or an equivalent media without serum.

4. Plasmids coding for the IL-1β bioluminescence sensor and for RLuc8 (*see* **Note 1**).

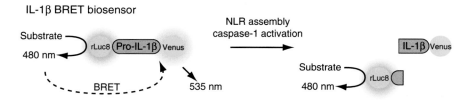

Fig. 1 Schematic representation of the IL-1β BRET biosensor. Upon caspase-1 activation, the BRET donor (rLuc8) and acceptor (Venus) molecules get apart, leading to a decrease of the net BRET signal

2.2 BRET Recording Using a Plate Reader

1. 96-well plate, white, flat and micro-clear bottom, cell culture treated, sterile with lids (*see* **Note 2**).

2. White backing tape to be stuck on the bottom of the 96-well plate the day of recording. It increases the luminescence signal by reflection.

3. A plate reader for luminescence recording equipped with two emission filters close to 480 nm and 535 nm.

4. HBS solution: 147 mM NaCl, 2 mM KCl, 2 mM $CaCl_2$, 1 mM $MgCl_2$, 10 mM HEPES, and 13 mM d-glucose, pH 7.4.

5. Multichannel pipettes.

6. Coelenterazine-h (*see* **Note 3**).

2.3 BRET Recording by Microscopy and Data Analysis

1. 35 mm dishes adapted for luminescence recording and treated for cell culture (e.g., we use μ-Dish from ibidi).

2. Coelenterazine-h (*see* **Note 3**).

3. Microscope and camera adapted for luminescence recording (e.g., our setup was the Olympus LV200 bioluminescence imaging system) and equipped with two emission filters close to 480 nm and 535 nm (*see* **Note 4**). To identify cells expressing the biosensor, filters for excitation and emission of Venus are also required (excitation close to 515 nm and emission close to 528 nm).

4. Software for image acquisition.

5. Software for data analysis, we use ImageJ (NIH), but equivalent software is also suitable.

3 Methods

Carry out all procedures with sterile and pyrogen-free material in biological safety cabinets Class II at room temperature unless otherwise specified.

3.1 Macrophage Transfection and Priming

1. Cells are plated 24 h before transfection. To use TransIT-Jurkat reagent, cells are plated to get 60–70 % confluence on day of transfection (*see* **Note 5**).

2. Transfection of macrophages can be performed following manufacturer instructions. Briefly, for one well of a 96-well plate, use 0.1 μg of DNA, 9 μl Opti-MEM, and 0.3 μl TransIT-Jurkat. Add DNA-transfection reagent complexes on well containing 50 μl of complete cell culture media.

3. Incubate cells with transfection mixture for 4–5 h at 37 °C and 5 % CO_2, then remove transfection reagent containing media, and replace it with fresh cell media.

4. BRET experiments can be performed the following day. If required, prime macrophages by adding 1 μg/ml of lipopolysaccharide to the culture media and incubated the cells for 4 h at 37 °C and 5 % CO_2.

3.2 Measuring IL-1β Processing in Real Time Using a Plate Reader

1. Pre-warm the plate reader at 37 °C.

2. Set up the plate reader controlling software according to your assay (*see* **Note 6**).

3. Predilute coelenterazine-h at 25 μM in HBS. Prepare 10 μl of solution per well to read (*see* **Note 7**).

4. Wash cells two times with 100 μl HBS per well and take care to not detach cells.

5. Keep cells in 40 μl of HBS per well.

6. Cell detachment can be checked using a transmitted light microscope.

7. Stick the white backing tape at the bottom of the plate.

8. Add 10 μl of prediluted coelenterazine-h per well to get a 5 μM final concentration.

9. Insert the plate in the plate reader.

10. Incubate for 6 min to allow the luminescence signal to reach a steady state (*see* **Note 8**).

11. Read the luminescence signal at 480 nm and 535 nm, and determine the gain according to the sensitivity of the plate reader (*see* **Note 9**). As a basal control, use nontransfected cells.

12. Initially use cells transfected with RLuc8 alone to record the luminescence signal at 480 nm and 535 nm in the same experimental conditions set above (*see* **Note 10**).

13. Then record luminescence signal at 480 nm and 535 nm from macrophages before any stimulation to determine the basal BRET signal.

14. BRET value, expressed in milliBRET units (or mBU), can be determined with the following equation:

$$\text{BRET (mBU)} = \left(\left(\frac{\text{Lum(535 nm)}}{\text{Lum(480 nm)}} \right)^{\text{IL 1β sensor}} - \left(\frac{\text{Lum(535 nm)}}{\text{Lum(480nm)}} \right)^{\text{RLuc 8 only}} \right) \times 1000$$

3.3 Monitoring Real-Time IL-1β Processing in Individual Macrophages

1. Pre-warm the microscope incubator at 37 °C.

2. Predilute coelenterazine-h at 200 μM in HBS. Prepare 100 μl of solution per well to read (*see* **Note 7**).

3. Wash the cells two times with 1 ml HBS and take care to not detach cells.

4. Keep cells in 0.9 ml of HBS per dish.

5. Place the dish in the microscope incubator and use appropriate objective to focus the cells.

6. Define the field and cells that you want to monitor.

7. Take a picture in bright field of your cells and after direct excitation of Venus (excitation close to 515 nm and emission close to 528 nm) to visualize cells that are transfected (*see* **Note 11**).

8. Gently add 100 μl of prediluted coelenterazine-h.

9. Incubate for 6 min to allow the luminescence signal to reach a steady state (*see* **Note 9**).

10. Sequentially record pictures at 480 nm and 535 nm, and stimulate the cells when required (*see* **Note 12**).

11. Export your pictures in tiff format for ImageJ analysis.

12. Using ImageJ software, define that division by 0 = 0. Go to Edit menu > options > misc, and in the "Divide by zero" window introduce the value "0.0" or "NaN".

13. Open the tiff pictures acquired at 480 nm and 535 nm.

14. Apply a median filter of 1 pixel for both pictures.

15. Remove background signal of the same area for both images. Define a field where no cells are present in the 480 nm picture and measure the maximum signal intensity (go to Analyze menu > Measure). Using the ROI manager tools (go to Analyze menu > Tools), repeat a similar measurement on the same field of the 535 nm pictures. Then subtract the correspondent value to the entire picture (go to Process menu > Math > Subtract).

16. Save the two pictures in tiff format with a different file name.

17. Divide the 535 nm pictures by the 480 nm pictures. Go to Process menu > Image calculator. Select the right picture for picture 1 and 2 and select "divide" from the drop-down menu.

18. BRET signals can be represented using a continuous 256-pseudocolor lookup table. Go to Image menu > Lookup tables and select the appropriate LUT.

4   Notes

1. EF-1a promoter is more appropriate than CMV promoter for protein expression in macrophages. Plasmid backbone size affects transfection efficiency (the smaller, the better).

2. 96-well plate with white bottom could be used but will not allow to visualize cells during the different steps of the experiment. BRET recording using a plate reader can be performed on other format plate (i.e., 384-well plate) depending on the assay and/or luminescence signal strength.

3. Coelenterazines are poorly soluble in water and must be resuspended in ethanol or methanol preparing a stock solution at 1 mM. Keep stock solution at –20 °C and protect it from light. For prolonged stability, resuspend coelenterazine in acidified ethanol (10 ml of ethanol and 200 µl of 3 N HCl) and store at –80 °C.

4. Any microscope can be used for BRET recording but will require an appropriate camera for bioluminescence detection and a light-tight enclosure.

5. If cells are less confluent or more confluent than 60–70 % on the day of transfection, it will increase cell death or reduce transfection efficiency, respectively. A cell line stably expressing the bioluminescence sensor can be also used to get more reproducible luminescence signal. Lentivirus can be used to produce such cell lines.

6. The detection of the donor and acceptor luminescence signal of a given well have to be recorded successively before switching to another well recording.

7. Prediluted solution of coelenterazine is susceptible to oxidation by air and thus should not be prepared in advance.

8. During the first 6 min following coelenterazine-h addition, luminescence signal will increase exponentially and might cause inappropriate BRET value (especially if the exposure time for each filter is ≥ 1 s). Thus, it is recommended to wait for the luminescence signal to reach a plateau.

9. Gain must be adjusted to get the widest dynamic range and the best sensitivity for a given plate reader. Alternatively, if possible, perform an automatic gain adjustment. These values of gain will stay approximately the same for each experiment and will just have to be adjusted depending on the transfection efficiency.

10. For a given set of filters, plate reader, and experimental conditions, the ratio $\left(\dfrac{\text{Lum(535nm)}}{\text{Lum(480nm)}} \right)^{\text{RLuc 8 only}}$ for the donor only willstay approximatively the same during each experiment and could be determined once.

11. Direct excitation of Venus using a laser after addition of coelenterazine-h results in a high background fluorescence. Always excite Venus before addition of the substrate.

12. As mentioned for the BRET recording using a plate reader, exposure time and gain amplification of signal have to be determined to get the best signal/background ratio.

Acknowledgments

This work was supported by the European Research Council and *Instituto de Salud Carlos III*-FEDER grants to P.P. V.C. was supported by the Institut National de la Santé et de la Recherche Médicale.

References

1. Medzhitov R (2008) Origin and physiological roles of inflammation. Nature 454(7203):428–435. doi:10.1038/nature07201
2. Dinarello CA (2009) Immunological and inflammatory functions of the interleukin-1 family. Annu Rev Immunol 27:519–550. doi:10.1146/annurev.immunol.021908.132612
3. Schroder K, Tschopp J (2010) The inflammasomes. Cell 140(6):821–832. doi:10.1016/j.cell.2010.01.040
4. Compan V, Baroja-Mazo A, Bragg L, Verkhratsky A, Perroy J, Pelegrin P (2012) A genetically encoded IL-1β bioluminescence resonance energy transfer sensor to monitor inflammasome activity. J Immunol 189:2131–2137. doi:10.4049/jimmunol.1201349

Chapter 6

Measuring IL-1β Processing by Bioluminescence Sensors II: The iGLuc System

Eva Bartok, Maria Kampes, and Veit Hornung

Abstract

Inflammasomes are multimeric protein complexes that proteolytically activate caspase-1, which subsequently matures cytokines of the IL-1 family and initiates the induction of pyroptotic cell death. Although this process is central both to pathogen defense and sterile inflammatory processes, there is currently no standard readout available for inflammasome activation which would be suitable for high-throughput applications. We have recently developed a new method for measuring inflammasome activation via the use of a novel proteolytic reporter iGLuc, an IL-1β Gaussia luciferase (iGLuc) fusion protein. Here, we provide detailed protocols for the use of iGLuc in transiently transfected or stably transduced cell lines. Using these protocols, IL-1β maturation as the result of inflammasome activation or other processes can be indirectly measured via the gain of Gaussia luciferase activity of cleaved iGLuc, allowing for rapid inflammasome reconstitution assays and high-throughput screening of inflammasome activity.

Key words Inflammasome, IL-1β, Gaussia luciferase, High-throughput screening

1 Introduction

Several NOD-like receptor proteins and also additional pattern recognition receptors are capable of forming multi-protein complexes, known as inflammasomes, with the adaptor protein ASC and caspase-1 (for a detailed overview, *see* 1 or 2). Inflammasome activation leads to the cleavage of the inflammatory cytokines IL-1β and IL-18 and the induction of an inflammatory form of cell death known as pyroptosis. At the same time, it has been shown that inflammasome activation is required for the release of a number of other proinflammatory factors, including IL-1α and HMGB1 [1–4]. All of these processes contribute decisively to the innate immune inflammatory response. In addition to inflammasome pathways that culminate in caspase-1 activation, several other inflammatory processes have been described as leading to the production of bioactive IL-1β. As such, it has been shown that caspase-8 can directly process IL-1β into its bioactive form that certain serine proteases can cleave IL-1β [5–7].

Francesco Di Virgilio and Pablo Pelegrín (eds.), *NLR Proteins: Methods and Protocols*, Methods in Molecular Biology, vol. 1417, DOI 10.1007/978-1-4939-3566-6_6, © Springer Science+Business Media New York 2016

Although all of the currently known inflammasomes seem to modulate the immune response to microbial agents, only AIM2 and NLRC4 have well-described interactions with ligands or accessory proteins. AIM2 directly binds to and is activated by cytosolic DNA [8], and the NLRC4 inflammasome is activated by bacterial flagellin and related T3SS needle proteins in an indirect process mediated by members of the NAIP family [9]. Although the NLRP6, NRLP7, and NLRP12 inflammasomes have been reported to be involved in microbial detection, no specific upstream activators of these sensors have been discovered so far [10–12].

In contrast, numerous upstream activators of the NLRP3 inflammasome have been reported, yet none of these seem to be direct NLRP3 ligands, and the process leading to NLRP3 activation is still not completely understood. Given its involvement in the pathogenesis of many widespread diseases, such as atherosclerosis, Alzheimer's disease, type II diabetes mellitus, and gout, NLRP3 has been the focus of intense research [13–16].

Thus, discovering new activators, inhibitors, and modulators for all of the inflammasomes would clearly be of interest both to clinicians and basic researchers alike. Unfortunately, research in this field has been slowed by the lack of a readout compatible with high-throughput screening (HTS) approaches. The gold standard for inflammasome activation is the measurement of caspase-1 activation via the immunoblotting of precipitated cellular supernatants. However, this approach is time-consuming, error prone, and unsuitable for HTS. In addition, this readout is inherently dependent on caspase-1 activation and thus not suitable for the measurement of other proteases capable of IL-1β processing (e.g., caspase-8, *see* above). Alternatively, approaches using ELISA for the detection of secreted IL-1β are possible in HTS-compatible systems, yet they require NF-κB-mediated upregulation of IL-1β, which may overshadow other important processes in inflammasome activation [17]. Here, we present a detailed protocol for a luciferase-based readout for IL-1β maturation, iGLuc, which we have recently developed [18]. iGLuc is based on a murine *IL-1β-Gaussia luc*iferase fusion protein and functions as a proteolytic reporter of inflammasome activation. In unstimulated cells, iGLuc demonstrates pro-IL-1β-dependent aggregation that abrogates the functionality of the C-terminal Gaussia luciferase tag. However, the activation of caspase-1 leads to the cleavage and monomerization of iGLuc, rendering the Gaussia luciferase moiety functional and inducing a robust luciferase signal [18].

The iGLuc reporter has a wide variety of possible uses in vivo and in vitro. In this protocol, we will concentrate on two important applications: (1) the transient reconstitution of the inflammasome with the iGLuc reporter, which can be used to investigate the functionality of different inflammasome proteins and activators of interest in a transient gain of function setting, and (2) the generation of lentivirally transduced cell lines, which stably express the iGLuc reporter. The former application can be used to investigate the comparative functionality of different inflammasome proteins, e.g., a comparison of a wild-type and a point-mutated variant of NLRP3, by making use of controlled inflammasome autoactivation through transient overexpression of its components. In addition, it is also possible to use this reconstitution system to test inflammasome activators and ligands. The latter approach (2) allows for the generation of stable, clonal cells with excellent signal to background. Although this readout is intrinsically dependent on the inflammasome pathways available in the transduced immune cells, their handling is relative simple, and the activation of such stable iGLuc cell lines is robust. Thus, stable iGLuc cell lines are particularly suited to HTS applications.

2 Materials

2.1 Inflammasome Reconstitution

1. Transfectable cell line: we recommend 293T-cell lines, such as 293FT from Invitrogen (*see* **Notes 1** and **2**).

2. Appropriate cell culture medium, e.g., Dulbecco's Modified Eagle's Medium (DMEM) supplemented with 10 % fetal calf serum (FCS) and 1 mM sodium pyruvate for 293T cells.

3. Serum-free cell culture medium suitable for transfection, e.g., Opti-MEM® (Life Technologies).

4. Transfection reagent with low cytotoxicity, e.g., GeneJuice (Novagen) (*see* **Note 3**).

5. Mammalian expression plasmids (*see* **Note 4**) containing:

   (a) The inflammasome protein of interest, e.g., wild-type human (hs) NLRP3 (*see* **Note 5**).

   (b) hs ASC with a C-terminal tag, such as HA (*see* **Note 6**).

   (c) hs caspase-1 (*see* **Note 7**).

   (d) iGLuc reporter.

6. A filler plasmid with a eukaryotic promoter (*see* **Note 8**).

7. 5× Passive lysis buffer (Promega) (*see* **Note 9**).

2.2 Generation of Stable iGLuc Cell Lines

2.2.1 Lentivirus Generation

1. 293T HEK cells for lentivirus generation (containing an SV40 large T antigen), such as 293FT from Life Technologies.

2. Appropriate cell culture medium (*see* **item 2**, Subheading 2.1).

3. 1.5 mg/mL poly-l-ornithine (p-l-ORN) 100× stock solution in water, filter-sterilized.

4. p-l-ORN 1× solution: 10 mL 100× stock diluted in 990 mL sterile deionized water.

5. 2× HBS buffer: 50 mM HEPES, 1.5 mM Na_2HPO_4, 280 mM NaCl, 10 mM KCl, 12 mM sucrose, filter-sterilized and with an optimized pH (*see* **Note 10**).

6. 2.5 M $CaCl_2$ filter-sterilized.

7. Sterile or purified deionized water.

8. Lentiviral plasmids:

   (a) Lentiviral expression plasmid containing iGLuc, e.g., pFUGW-iGLuc (*see* **Note 11**).

   (b) Lentiviral packaging plasmid(s) of the appropriate generation, e.g., pCMV-dR8.2 (Weinberg lab; Addgene #8455) or psPax2 (Trono Lab; Addgene #12259) (*see* **Note 12**).

   (c) Lentiviral envelope plasmid, e.g., pCMV-VSV-G (Weinberg lab; Addgene #8454) (*see* **Note 13**).

9. 0.45 μm polyethersulfone (PES)-filters or filter flasks.

10. Sterile 20 mL syringes.

11. For the ultracentrifuge:

    (a) Tubes (*see* **Note 14**).

    (b) 70 % ethanol.

    (c) 20 % sucrose solution.

2.2.2 Transduction

1. Transduceable cell line, e.g., immortalized murine macrophages, THP-1, U937, etc. (*see* **Note 15**).

2. 8 mg/mL polybrene solution in sterile deionized water (*see* **Note 16**).

2.3 Gaussia Luciferase Assay

1. White, opaque luminescence plate in 96-well or 384-well format.

2. 1 mg/mL coelenterazine stock solution prepared in 100 % ethanol. Store solution at –20 °C.

3. Sterile, deionized water.

3 Methods

3.1 Inflammasome Reconstitution

3.1.1 Plating 293T Cells

1. Plate 100 μL of 293T cells at 3×10^6 cells/mL in a 96-well plate using pre-warmed cell culture medium (37 °C).

   (a) For the first experiment, one should plan enough conditions to titrate all of the inflammasome proteins, e.g., caspase-1 and ASC, whatever upstream NLR is being used. We do not recommend using a ligand or activator at this stage (*see* **Note 17**).

   (b) There should be enough conditions to allow for triplicates to compensate for the inherent technical error in luciferase-based assays.

2. Allow the cells to settle in the well for 5–10 min at RT before placing cells in a cell culture incubator.

3. Wait at least 5 h for the cells to become adherent and spread out in the wells (*see* **Note 18**).

3.1.2 Plasmid Transfection

1. In the meantime, prepare the plasmids for transfection (*see* **Note 19**). A total plasmid weight of 212.5 ng will be used per 96-well.

2. iGLuc will be used at a constant amount of 100 ng per well.

3. For the first experiment, we recommend performing titrations in serial dilutions as follows (*see* **Notes 20** and **21** and Fig. 1):

   (a) The NLR of interest (e.g., NLRP3): 75–0 ng.

   (b) ASC: 12–0 ng.

   (c) Caspase-1: 25–0 ng.

   (d) Filler plasmid: as necessary for a final amount of 212 ng.

4. Plasmids should be diluted into Opti-MEM (*see* **item 3**, Subheading 2.1) at a total concentration of 212.5 ng plasmid per 25 μL medium (*see* **Note 22**).

5. We recommend dispersing the different plasmid combinations on a 96-well plate in the same layout as later used for the cell transfection. Please *see* Fig. 1 for a concrete example of a pipetting plan.

6. If a ligand or activator is being tested, then an additional level of titration is necessary for the ligand itself.

7. Prepare the transfection reagent (*see* **item 4**, Subheading 2.1): calculate using 0.5 μL GeneJuice and 25 μL Opti-MEM per well with an additional buffer for pipetting loss (*see* **Note 22**).

8. Combine the GeneJuice and Opti-MEM in a 12-channel reservoir using one reservoir channel for each row on a 96-well plate. For the plate outline shown in Fig. 1g, this would mean using five slots of the reservoir (B, C, D, E, F). After adding the transfection reagent, pipette the mixture gently up and down several times using a 1 mL pipette.

9. Afterward, wait 5 min to allow the reagent to disperse in the Opti-MEM.

10. Pipette the appropriate volume of GeneJuice/Opti-MEM mix onto each well containing plasmid/Opti-MEM mix using a multichannel pipette. The ratio should be 1:1 between the plasmid mix and the added GeneJuice mix.

11. Allow the GeneJuice/plasmid mix to incubate for 20 min at RT.

12. Transfer 50 μL of GeneJuice/plasmid mix per well to the 293T-cell culture plates using a multichannel pipette.

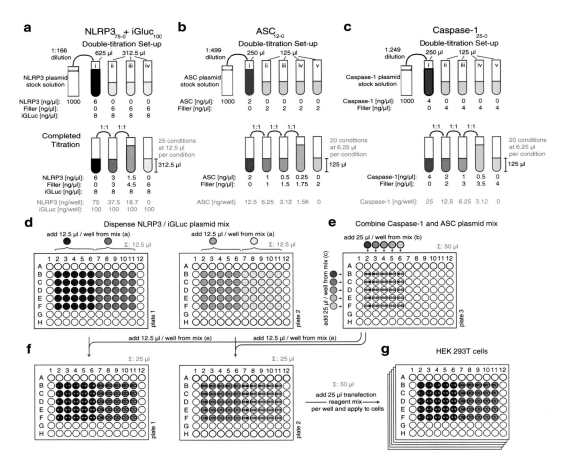

Fig. 1 Pipetting scheme for a HEK293T iGLuc reporter experiment. A pipetting scheme is depicted for an experiment in which NLRP3, ASC, and caspase-1 are titrated over the following concentrations: NLRP3 (75–0 ng/well; three serial dilutions), ASC (12.5–0 ng/well; four serial dilutions), and caspase-1 (25–0 ng/well; four serial dilutions). These and the iGLuc reporter plasmid are combined to a total of 212.5 ng transection mix per 96-well. Generating all possible combinations of components yield a total of 100 different conditions ($4_{NLRP3} \times 5_{ASC} \times 5_{caspase-1} = 100$). One approach to creating the necessary plasmid mix is the use of a serial dilution, which is shown for NLRP3, ASC, and caspase-1 (**a–c**). (**a**) The iGLuc reporter plasmid is included in the NLRP3 titration mix to obtain a final amount of 100 ng per well. Opti-MEM containing 8 ng/μL iGLuc plasmid (= 100 ng per 12.5 μL) is used for the serial dilutions of the NLRP3 plasmid. NLRP3 is added to obtain a concentration of 6 ng/μL in the first vial (1) containing 625 μL Opti-MEM. From this mixture, 50 % (= 312.5 μL) is added to the next vial (2) containing 6 ng/μL of a stuffer plasmid in a volume of 312.5 μL. This step is repeated once into a third vial (3), whereas the fourth vial (4) only contains the stuffer plasmid at 6 ng/μL. Finally, this creates four vials containing a total volume of 312.5 μL Opti-MEM with 75 (1), 37.5 (2), 18.7 (3), and 0 (4) ng NLRP3 plasmid and the required amount of stuffer plasmid for each transfection condition (volume of 12.5 μL at this point). See *pink* label at the *bottom* of the panel. (**b**) and (**c**), analogous dilutions are performed for ASC and caspase-1, starting instead with a volume of 250 μL and a concentration of 2 ng/μL ASC or 4 ng/μL caspase-1 in the first vial (1) and the equivalent amounts of the stuffer plasmids in the vials to be diluted in (2–5). 125 μL is then transferred from the first into the next vial containing 125 μL, and the serial dilution is then continued to obtain a total of five vials. The respective amounts of ASC (**b**) and caspase-1 (**c**) plasmid per transfection condition (volume of 6.25 μL at this point) are depicted at the bottom. (**d**) 12.5 μL of each of the four vials of the NLRP3/iGLuc mix obtained in **a** is then dispensed into 25 (5 × 5) wells of a 96-well plate,

13. Immediately place the cell culture plates back into the incubator, and allow approximately 20 h for plasmid expression for functional tests of inflammasome proteins. Allow 16 h before stimulation for experiments involving ligands or activators (*see* **Note 23**).

*3.1.3   Cell Lysis
and Luciferase Readout*

1. Dilute 5× passive lysis buffer in a sterile, deionized water.

2. Remove the cell culture plates from the incubator, and carefully remove the cell culture medium from the wells (*see* **Note 24**).

3. Add 30 μL passive lysis buffer to each well, and incubate the plates on a shaker for 10 min at RT (*see* **Note 25**).

4. When the cells are lysed, remove 15 μL of the cell/lysis buffer mix per well to be transferred to a luminescence reader plate.

5. Dilute the coelenterazine stock 1:500 in a sterile, deionized water.

6. Add 15 μL of diluted coelenterazine solution to the 15 μL of cell lysate in each well (final coelenterazine concentration = 1 μg/mL).

7. Rock the plate back and forth gently but quickly. Then, place it into the luminescence reader to measure the relative light units (RLU).

8. The remaining cell lysate can be combined with 2× Laemmli buffer for immunoblotting if desired. This can be helpful for troubleshooting (*see* **Note 26**).

3.1.4   Data Analysis

1. Many of the conditions will have high RLU counts (*see* Fig. 2a, b). Conditions using fixed amounts of ASC and caspase-1 that lead to autoactivation without the addition of an NLR should be discarded for further studies, in which an upstream inflammasome activator is being tested (*see* Fig. 2c).

2. If the functionality of an inflammasome protein is being tested, then the conditions of interest are those where uniquely the titration of the NLR leads to inflammasome activation (*see* Fig. 2c).

Fig. 1 (continued) thus yielding a total number of 100 different wells. (**e**) The ASC and the caspase-1 plasmid mixes obtained in (**b**) and (**c**) are then orthogonally mixed to obtain a total of 25 different plasmid combinations by combing 25 μL of each mix to a total volume of 50 μL. (**f**) of the different mixes obtained in (**e**) 12.5 μL per mix are transferred to the pre-aliquoted NLRP3/iGLuc mixes (**d**) to obtain a total volume of 25 μL. These plasmid mixes are then combined with 25 μL Opti-MEM containing 0.5 μL GeneJuice. After an additional incubation period of 20 min, the 50 μL transfection mix is then added to 293T cells cultured in 96-well plates (**g**). Of note, the volumes described here do not account for pipetting losses and do not take into consideration that duplicates or triplicates should be run. As such, the volumes should be multiplied by the number of replicates planned. We recommend approximately 10 % surplus for the serial dilutions (**a–c**) and for the subsequent mixing steps (**d–f**)

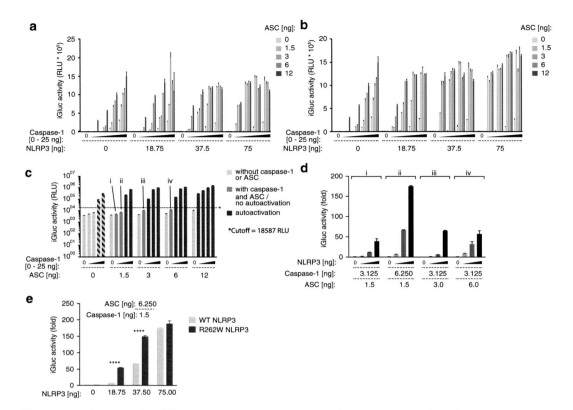

Fig. 2 Reconstitution of the NLRP3 inflammasome employing the iGLuc reporter in HEK293T cells. HEKs were transfected as indicated. Data depicted in (**a**) and (**b**) include all data points for WT NLRP3 (**a**) and R260W NLRP3 (**b**). In order to compare NLRP3 to NLRP3 R260W as in (**e**), conditions without autoactivation (**c**) but with a good signal to background (**d**) have to be determined. In (**c**), the base-level activation of the inflammasome without NRLP3 is examined for all caspase-1/ASC combinations. The conditions without ASC or caspase-1 (*light gray*) or autoactivation (*dark gray*) should be excluded. Autoactivation is defined as $\mu + 5\sigma$. The conditions suitable for further analysis are shown in *blue*. In (**d**), the signal-to-background ratio following WT NLRP3 titration is shown for the conditions with ASC and caspase-1 and without autoactivation (*blue*) in (**c**). Condition (2) has the best signal to background and is used for subsequent comparison with R260W NLRP3 in (**e**)

3. If several conditions meet this criterion, a titration condition can be selected according to its signal to background (*see* Fig. 2d).

4. Comparisons between two NLRs should then be made with fixed ASC and caspase-1 concentrations (*see* Fig. 2e).

5. If a ligand or activator is being tested, then only fixed amounts of caspase-1, ASC, and the NLR, which additionally require the ligand for induction of luciferase activity, are of interest.

3.2 Generation of Stable iGLuc Cell Lines

3.2.1 Plating Cells for Lentivirus Production

1. Coat 10 cm tissue culture dishes with 5 mL of 1× p-l-ORN per dish, and incubate either at 4 °C overnight or at 37 °C for 2 h.

2. Wash the dishes at least 3 times using generous amounts of PBS.

3. Using pre-warmed medium at 37 °C, plate 1×10^7 cells per 10 cm dish in 10 mL cell culture medium.

4. Allow the cells to settle for 5–10 min at RT.

5. Place the cells in the incubator for at least 5 h.

6. Change the medium to 8 mL of new, pre-warmed DMEM just before completing the $CaPO_4$ transfection (*see* **Note 27**).

3.2.2   Calcium Phosphate Transfection

1. Combine the lentiviral plasmids in an Eppendorf or Falcon tube, using the following amounts per 10 cm dish (*see* **Note 28**).

   (a) 20 μg iGLuc lentiviral plasmid (e.g., pFUGW-iGLuc).

   (b) 13.5 μg lentiviral packaging plasmid (e.g., pCMV-dR8.2).

   (c) 6.5 μg lentiviral envelope plasmid (e.g., pCMV-VSV-G).

2. Use sterile, deionized water to fill the plasmid mix to a volume of 500 μL.

3. Add 500 μL 2× HBS solution.

4. Vortex the Eppendorf tube or Falcon gently under the bench.

5. While vortexing, add 50 μL 2.5 M $CaCl_2$ (*see* **Note 29**).

6. Continue vortexing for another 10–15 s.

7. Allow the $CaPO_4$ plasmid solution to incubate for approximately 25 min at RT.

8. Pipette the solution onto the 293T cells while gently rocking the tissue culture dish, so that the $CaPO_4$ solution is distributed evenly throughout the medium.

9. After 8–12 h, change the medium again. The supernatants can be harvested after another 24 h.

3.2.3   Harvesting Viral Supernatants

1. Carefully harvest the supernatant from the tissue culture dish and place in 50 mL Falcon tubes. If desired, new medium can be added to the plate and harvested after another 24 h.

2. Centrifuge the supernatant at $1000 \times g$ for 10 min to precipitate cellular debris from the 293T cells.

3. Transfer the supernatant to a new Falcon tube.

4. Use a 0.45 μm filter or filter flask to remove any remaining cellular debris.

3.2.4   Ultracentrifugation of Viral Supernatants (Optional, See Note 30)

1. Rinse ultracentrifugation tubes with 70 % and leave to dry.

2. Place the tubes into the ultracentrifuge buckets.

3. For a Beckman Coulter SW32 Rotor:

   (a) Load 34 mL of the viral supernatant per tube. Alternatively, 28 mL supernatant with a 4 mL sucrose cushion can be used (see Note 30).

   (b) Tare the buckets carefully, with a maximum weight difference of 0.02 g for each bucket pair.

   (c) Run the ultracentrifuge at 4 °C for 2 h at approximately $100,000 \times g$ (max), which corresponds to 25000 rpm using a Beckman Coulter SW32 Rotor.

(d) Decant the supernatant and add the desired amount of PBS to the pellet.

(e) To dissolve the pellet, incubate the tube for 2 h at 4 °C using a gentle shaker.

3.2.5 Storage of Lentivirus

1. After filtration or concentration, the lentivirus can be stored at −80 °C until their further use.

3.2.6 Viral Spin Transduction

1. In a 12-well plate, plate 3×10^5 of target cells (e.g., immortalized murine macrophages) per well in 500 μL of cell culture medium.

    (a) For adherent cells, allow cells to settle for 5–10 min at RT. Then, place them in the incubator for 1–2 h (until they are adherent).

    (b) For suspension cells, continue directly to **step 2**.

2. Add polybrene to the wells with a final concentration of 4–8 g/mL (*see* **Note 16**).

3. Add lentivirus at MOI 1–10 to the wells (*see* **Note 31**).

4. Spin transduce at $600–800 \times g$ for 45–90 min at 32 °C using a swing-bucket centrifuge.

5. Place cells back in the incubator for at least 2 days.

3.2.7 Production and of Monoclonal Reporter Cell Lines

1. Plate cells in 96-well plates using limited dilution with a calculated 0.3 cell per well (*see* **Note 32**).

2. Allow the single-cell clones to expand. This may take 2–3 weeks depending on the cell type.

3. When cells have sufficiently expanded, perform an inflammasome activation assay with several standard ligands to determine which clones should be kept in culture for future assays.

3.2.8 iGLuc Assay in Stable Cells for Clone Selection

1. Plate cells in a 96-well format (*see* **Note 33**).

    (a) Immortalized mouse macrophages: 1×10^5 cells per well.

    (b) PMA-differentiated THP-1 cells: 7×10^4 cells per well.

2. Test the cell lines with standard inflammasome stimuli, e.g.:

    (a) 5 μM nigericin as a NLRP3 stimulus following a 4 h priming phase with an NF-kB inductor such as LPS (*see* **Note 34**).

    (b) 200 ng plasmid DNA as an AIM2 stimulus.

    (c) Transfected flagellin or salmonella infection as a NLRC4 stimulus (*see* **Note 35**).

3. After stimulation (*see* **Note 36**), remove 25 μL cellular supernatant per condition and transfer the supernatant to a white luminescence assay plate.

4. Dilute coelenterazine stock 1:500 in water (*see* **Note 37**).

5. Add 25 μL diluted coelenterazine solution per well containing cellular supernatant.

6. Rock the plate back and forth gently but quickly, and place the plate in the reader.

3.2.9   Data Analysis

1. Clones should be selected according to two main criteria (*see* Fig. 3):

   (a) The clones should have a good signal-to-background ratio. Here, it is important to calculate the normalized fold iGLuc induction. For mouse macrophages, this should be more than 1:50 during the initial screening of the clones (*see* Fig. 3a, b).

   (b) The clones should respond appropriately to inflammasome stimuli in accordance with their genotype (*see* Fig. 3a).

2. It is recommended to keep at least five to ten good clones when planning HTS experiments (*see* **Note 38**).

4   Notes

1. Theoretically, it is possible to use any cell line that is permissible to transfection for these experiments. However, constitutive expression of inflammasome components within the cell line must be taken into account, e.g., caspase-1 expression in HeLa cells has been reported [19, 20], since this may disturb the assay.

2. The cells used should be split regularly and not be allowed to become confluent. Cells that are in adequate condition to express a simple GFP construct may still not be able to take up several different plasmids. *This is a very common problem for those new to this assay.*

3. We have had good results using GeneJuice, and our protocols have been optimized for this reagent. However, there are certainly alternative transfection reagents, which would also work with this protocol.

4. We have generally used plasmids with strong promoters (EF1a, CMV), and our protocols have been optimized for these constructs. Other promoters can be used. However, the plasmid amounts and the time allowed for protein expression will probably have to be optimized. Of note, the expression kinetics are particularly difficult to control if promoters are mixed.

5. It is also possible to use inflammasome proteins from other species. So far, we have experienced no difficulties combining murine and human inflammasome proteins, e.g., hs ASC with

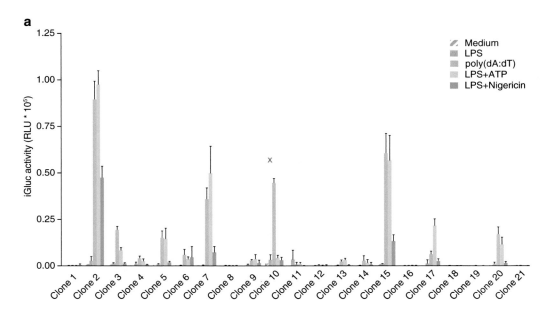

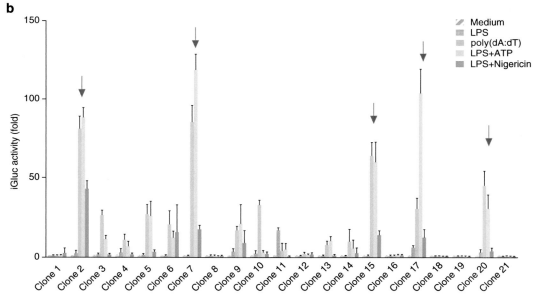

Fig. 3 Selection of iGLuc-transduced macrophage cell clones. (**a**) Murine immortalized macrophages were transduced using the iGLuc reporter plasmid, and single-cell clones were generated using limiting dilution cloning. Clones were seeded, primed with LPS for 4 h, and then stimulated as indicated. The supernatants were analyzed after 6 h for iGLuc activity. Here, the absolute counts (RLU) are shown. As indicated with the *red x*, not all clones react to all stimuli. (**b**) Evaluation of the signal-to-background ratio (dataset as in **a**). Shown is the fold induction of iGLuc activity normalized to unstimulated cells. The clones that were selected for further use are indicated with *red arrows*. Please note that some clones with lower absolute RLUs still demonstrated excellent signal-to-background ratios

mm NLRP3. However, we would still recommend using one species if a novel protein is being tested.

6. ASC, all of the NLRs we have tested, and AIM2 tolerate C-terminal tags. Using tags facilitates Western blotting if you suspect protein expression problems. In addition, using ASC-GFP in the assay allows the use of ASC specking as a simultaneous microscope-based readout.

7. Although we have tried several C-terminal tags on our caspase-1 constructs, we have still never been able to detect caspase-1 in 293T cells via immunoblotting for a C-terminal tag. Caspase-1 is clearly expressed in 293T cells as evidenced by the observed IL-1β cleavage after caspase-1 overexpression. We do not have a definitive explanation for this phenomenon, but it is possible that the half-life of the tagged caspase-1 construct is too short to be measured via immunoblotting.

8. An empty plasmid or one expressing a protein irrelevant for inflammasome activation, e.g., GFP, can be used. One advantage to using a fluorescent protein is that it can be used to roughly determine transfection efficiency.

9. We have had the best results with this product so far, yet other lysis buffers might be suitable as well. Of note, 0.5 % saponin-based buffers, which are commonly used for firefly luciferase assays do not work well with the iGLuc system.

10. A good 2× HBS buffer is essential for efficient transduction. There are many different recipes for 2× HBS, and we have included one of the several recipes used in our lab, originally taken from [21]. Here, the most important step seems to be pH$^+$ optimization. In keeping with Kutner et al., we recommend creating three stocks of 2× HBS by adjusting to different pH$^+$ levels: 7.07, 7.09, and 7.11. The three different solutions should then be tested for their transfection efficiency, and the best solution should be stored for further use. Another 2× HBS recipe and production protocol is available at http://www.lentiweb.com.

11. pFUGW is a small lentiviral construct without a selection marker, yet it yields very high titers. To enrich iGLuc-expressing cells, a lentivirus with a selection marker should be used (*see* **Note 31**).

12. Please *see* https://www.addgene.org/lentiviral/packaging/ for more details on lentiviral packaging.

13. Pseudotyping using VSV-G ensures a wide tropism; however, depending on the target cell type, the use of other envelopes is of course also possible.

14. Each ultracentrifuge rotor has its own tube specifications. We use the ultra-clear tubes from Beckman Coulter for the Beckman Coulter rotor SW 32 Ti.

15. 293T HEK cells can be transduced with unconcentrated virus and immortalized macrophages as well, if the produced titer is high enough. Many other cells, such as THP-1, often require concentrated virus and spin transduction. New protocols must be established for every cell line and application.

16. We use 8 µg/mL polybrene to transduce THP-1 cells. Whether polybrene should be used and in what amount has to be optimized for each cell line. One alternative is protamine [22].

17. For initial experiments, many conditions will be necessary. For example, for a four-step titration of WT NLRP3 and five-step titration of ASC and caspase-1, 100 conditions are required. A comparison of WT to R262W NLRP3—a mutation associated with Muckle-Wells disease [23]—200 conditions are required. Adding a titrated upstream activator at this stage would obviously go beyond the scope of an initial experiment. As such, we recommend an iterative approach to determine the optimal transfection conditions with a favorable activation "window" for subsequent experiments.

18. As in **Note 2**, it is very important that your cells are "healthy looking." Before transfection, the cells should be evenly distributed in the plate and have spread out with extensions.

19. It may be helpful to review the model pipetting scheme in Fig. 1 before planning the experiment.

20. We recommend using at least four titration steps. However, it may be necessary to readjust the titration scale for a second experiment if too much autoactivation is seen. In Fig. 1, five steps are used for caspase-1 and ASC and four for NRLP3.

21. It is possible to titrate the inflammasome component-encoding plasmid DNA directly into Opti-MEM containing the diluted filler plasmid. We call this "double titration," and it saves the experimenter a considerable amount of time. Please *see* Fig. 1a–c for an illustration.

22. Please round up these volumes to compensate for pipette-dependent volume loss, e.g., 10 %.

23. This parameter strongly depends on the promoters used. Twenty hours is a good starting point for CMV promoters. EF1a promoters may need a few hours longer. Other promoters may require some optimization before such a large-scale experiment can be done.

24. Leaving culture medium in the well is a common source of error since the remaining medium causes background in the

luciferase readout. As such, it is important to meticulously remove the medium left in the well.

25. Of note, active IL-1β is not excreted from HEK 293T cells (this might by different for other transfectable cell lines, e.g., HeLa or HaCaT cells). Thus, HEK 293T cells must be lysed before the iGLuc readout. We recommend using Promega 5× passive lysis buffer for lysis (*see* **Note 9**). The lysis conditions for different cell lines may vary, so please verify lysis using a light microscope.

26. If the reconstitution does not work properly, it is important to verify that all of the proteins are being expressed and that their expression is at the expected level.

27. This step should not be overlooked. The acidic metabolic products found in the medium (even after a few hours) lower transfection efficiency.

28. These are reference values and should be optimized for your system.

29. There are many ways of approaching this step. Some experimenters prefer vortexing after adding the $CaCl_2$ since the handling is easier. Others invert the tube. It is best to try out several protocols and see what works for you.

30. Different cell lines require different titers for transduction (*see* **Note 15**). Thus, concentrating the lentivirus is not always necessary. In addition, using a sucrose cushion is only necessary for cell lines that are particularly sensitive to viral debris or FCS. For a good protocol using 20 % sucrose, *see* Kutner et al. There is also a detailed alternative protocol for lentivirus precipitation using PEG6000 in the manuscript.

31. There are many different protocols for determining the number of transducing units per mL. Due to its ease of use, we recommend p24 ELISA.

32. We generally use pFUGW, which does not encode for a selection marker, and we have had good results using limited dilution without prior selection.

33. For other cell lines, we recommend to establish the cell stimulation protocol beforehand using standard inflammasome stimuli and IL-1β or IL-18 ELISA as a readout.

34. For mouse macrophages, we use 200 ng/mL LPS for 4 h to prime NLRP3 expression, yet it is also possible to use other TLR ligands.

35. This must be carefully established beforehand in non-transduced cells. Flagellin can be transfected with DOTAP, but this must be optimized for every cell type. A stimulation protocol using live salmonella can be found in [24].

36. iGLuc is excreted from macrophage and monocyte cell lines and can thus be measured directly from the supernatant. The amount of time needed for this process depends on the cell line used. For immortalized macrophages, we generally wait at least 4 h before measuring iGLuc activity since AIM2 activation upon plasmid DNA transfection needs 4–6 h. Please *see* [24] for a protocol.

37. For HTS applications, it may be preferable to use a buffer, which stabilizes Gaussia luciferase activity. For a good buffer recipe, please *see* [25].

38. Here, it is important to consider clonal effects and to try to avoid such biases in a screening. Optimally, it would be best to use more than one clone for a library screening.

References

1. Bauernfeind F, Ablasser A, Bartok E, Kim S, Schmid-Burgk J, Cavlar T, Hornung V (2010) Inflammasomes: current understanding and open questions. Cell Mol Life Sci 68:765–783

2. Latz E, Xiao TS, Stutz A (2013) Activation and regulation of the inflammasomes. Nat Rev Immunol 13:397–411

3. Bauernfeind F, Hornung V (2013) Of inflammasomes and pathogens—sensing of microbes by the inflammasome. EMBO Mol Med 5:814–826

4. Groß O, Yazdi AS, Thomas CJ, Masin M, Heinz LX, Guarda G, Quadroni M, Drexler SK, Tschopp J (2012) Inflammasome activators induce interleukin-1α secretion via distinct pathways with differential requirement for the protease function of caspase-1. Immunity 36:388–400

5. Gringhuis SI, Kaptein TM, Wevers BA, Theelen B, van der Vlist M, Boekhout T, Geijtenbeek TB (2012) Dectin-1 is an extracellular pathogen sensor for the induction and processing of IL-1beta via a noncanonical caspase-8 inflammasome. Nat Immunol 13(3):246–254. doi:10.1038/ni.2222

6. Bossaller L, Chiang PI, Schmidt-Lauber C, Ganesan S, Kaiser WJ, Rathinam VA, Mocarski ES, Subramanian D, Green DR, Silverman N, Fitzgerald KA, Marshak-Rothstein A, Latz E (2012) Cutting edge: FAS (CD95) mediates noncanonical IL-1beta and IL-18 maturation via caspase-8 in an RIP3-independent manner. J Immunol 189(12):5508–5512. doi:10.4049/jimmunol.1202121

7. Netea MG, van de Veerdonk FL, van der Meer JW, Dinarello CA, Joosten LA (2014) Inflammasome-independent regulation of IL-1-family cytokines. Annu Rev Immunol. doi:10.1146/annurev-immunol-032414-112306

8. Jin T, Perry A, Jiang J, Smith P, Curry JA, Unterholzner L, Jiang Z, Horvath G, Rathinam VA, Johnstone RW, Hornung V, Latz E, Bowie AG, Fitzgerald KA, Xiao TS (2012) Structures of the HIN domain:DNA complexes reveal ligand binding and activation mechanisms of the AIM2 inflammasome and IFI16 receptor. Immunity 36(4):561–571. doi:10.1016/j.immuni.2012.02.014

9. von Moltke J, Ayres JS, Kofoed EM, Chavarria-Smith J, Vance RE (2013) Recognition of bacteria by inflammasomes. Annu Rev Immunol 31:73–106. doi:10.1146/annurev-immunol-032712-095944

10. Elinav E, Strowig T, Kau AL, Henao-Mejia J, Thaiss CA, Booth CJ, Peaper DR, Bertin J, Eisenbarth SC, Gordon JI (2011) NLRP6 inflammasome regulates colonic microbial ecology and risk for colitis. Cell 145:745–757

11. Khare S, Dorfleutner A, Bryan NB, Yun C, Radian AD, de Almeida L, Rojanasakul Y, Stehlik C (2012) An NLRP7-containing inflammasome mediates recognition of microbial lipopeptides in human macrophages. Immunity 36:464–476

12. Vladimer GI, Weng D, Paquette SWM, Vanaja SK, Rathinam VAK, Aune MH, Conlon JE, Burbage JJ, Proulx MK, Liu Q, Reed G, Mecsas JC, Iwakura Y, Bertin J, Goguen JD, Fitzgerald KA, Lien E (2012) The NLRP12 inflammasome recognizes Yersinia pestis. Immunity 37:96–107

13. Duewell P, Kono H, Rayner KJ, Sirois CM, Vladimer G, Bauernfeind FG, Abela GS,

Franchi L, Núñez G, Schnurr M, Espevik T, Lien E, Fitzgerald KA, Rock KL, Moore KJ, Wright SD, Hornung V, Latz E (2010) NLRP3 inflammasomes are required for atherogenesis and activated by cholesterol crystals. Nature 464:1357–1361

14. Heneka MT, Kummer MP, Stutz A, Delekate A, Schwartz S, Vieira-Saecker A, Griep A, Axt D, Remus A, Tzeng TC, Gelpi E, Halle A, Korte M, Latz E, Golenbock DT (2013) NLRP3 is activated in Alzheimer's disease and contributes to pathology in APP/PS1 mice. Nature 493 (7434):674–678

15. Masters SL, Dunne A, Subramanian SL, Hull RL, Tannahill GM, Sharp FA, Becker C, Franchi L, Yoshihara E, Chen Z, Mullooly N, Mielke LA, Harris J, Coll RC, Mills KHG, Mok KH, Newsholme P, Núñez G, Yodoi J, Kahn SE, Lavelle EC, O'Neill LAJ (2010) Activation of the NLRP3 inflammasome by islet amyloid polypeptide provides a mechanism for enhanced IL-1[beta] in type 2 diabetes. Nat Immunol 11:897–904

16. Martinon F, Pétrilli V, Mayor A, Tardivel A, Tschopp J (2006) Gout-associated uric acid crystals activate the NALP3 inflammasome. Nature 440:237–241

17. Bauernfeind FG, Horvath G, Stutz A, Alnemri ES, MacDonald K, Speert D, Fernandes-Alnemri T, Wu J, Monks BG, Fitzgerald KA, Hornung V, Latz E (2009) Cutting edge: NF-kappaB activating pattern recognition and cytokine receptors license NLRP3 inflammasome activation by regulating NLRP3 expression. J Immunol 183:787–791

18. Bartok E, Bauernfeind F, Khaminets MG, Jakobs C, Monks B, Fitzgerald KA, Latz E, Hornung V (2013) iGLuc: a luciferase-based inflammasome and protease activity reporter. Nat Methods 10:147–154

19. Abdul-Sater AA, Koo E, Hacker G, Ojcius DM (2009) Inflammasome-dependent caspase-1 activation in cervical epithelial cells stimulates growth of the intracellular pathogen chlamydia trachomatis. J Biol Chem 284:26789–26796

20. Cheng W, Shivshankar P, Li Z, Chen L, Yeh I-T, Zhong G (2008) Caspase-1 contributes to *Chlamydia trachomatis*-induced upper urogenital tract inflammatory pathologies without affecting the course of infection. Infect Immun 76:515–522

21. Kutner RH, Zhang X-Y, Reiser J (2009) Production, concentration and titration of pseudotyped HIV-1-based lentiviral vectors. Nat Protoc 4:495–505

22. Cornetta K, Anderson WF (1989) Protamine sulfate as an effective alternative to polybrene in retroviral-mediated gene-transfer: implications for human gene therapy. J Virol Methods 23:187–194

23. Dode C, Le Du N, Cuisset L, Letourneur F, Berthelot JM, Vaudour G, Meyrier A, Watts RA, Scott DG, Nicholls A, Granel B, Frances C, Garcier F, Edery P, Boulinguez S, Domergues JP, Delpech M, Grateau G (2002) New mutations of CIAS1 that are responsible for Muckle-Wells syndrome and familial cold urticaria: a novel mutation underlies both syndromes. Am J Hum Genet 70(6):1498–1506

24. Kim S, Bauernfeind F, Ablasser A, Hartmann G, Fitzgerald KA, Latz E, Hornung V (2010) Listeria monocytogenes is sensed by the NLRP3 and AIM2 inflammasome. Eur J Immunol 40:1545–1551

25. Wu C, Suzuki-Ogoh C, Ohmiya Y (2007) Dual-reporter assay using two secreted luciferase genes. Biotechniques 42:290, 292

Chapter 7

Assessing Extracellular ATP as Danger Signal In Vivo: The pmeLuc System

Francesco Di Virgilio, Paolo Pinton, and Simonetta Falzoni

Abstract

Inflammation is the key pathophysiological response triggered by noxious agents in multicellular organisms. Central to inflammation is detection of exogenous or endogenous danger signals by immune cells. Extracellular ATP is a ubiquitous danger signal released during septic or sterile inflammation. The development of reliable techniques to measure extracellular ATP in vivo has become an urgent need in inflammation studies after the discovery that the most potent plasma membrane receptor responsible for NLRP3 inflammasome activation is an ATP-activated receptor, P2RX7. Here we describe an easy bioluminescence technique for the measurement of extracellular ATP in vivo.

Key words Extracellular ATP, Plasma membrane luciferase, Luciferin, Luminescence, Luminometry

1 Introduction

Bioluminescence is a natural phenomenon, due to chemical emission of light (chemiluminescence) by living organisms, conserved in many different species (bacteria, protists, fungi, insect, several marine organisms) with the notable exception of higher terrestrial organisms. This process yields photons as a consequence of an exergonic reaction catalyzed by a family of enzymes (e.g., luciferases) that oxidize a photon-emitting substrate. Luciferase (Luc), mainly derived from the firefly *Photinus pyralis* or from the coelenterate *Renilla reniformis*, has been long used as an in vitro assay to measure ATP [1]. Firefly luciferase is a 62 kDa protein belonging to the adenylating enzyme superfamily. In the presence of magnesium ions, molecular oxygen, and ATP, luciferase catalyzes oxidation of the substrate D-luciferin (LH_2) accompanied by light emission in the green to yellow region ($\lambda max = 560$ nm). The reaction proceeds through three main steps:

Francesco Di Virgilio and Pablo Pelegrín (eds.), *NLR Proteins: Methods and Protocols*, Methods in Molecular Biology, vol. 1417, DOI 10.1007/978-1-4939-3566-6_7, © Springer Science+Business Media New York 2016

1. Formation of the intermediate Luc-D-luciferyl adenylate (LH_2-AMP), with release of inorganic phosphate (PPi):

$$Luc + LH_2 + ATP \rightarrow Luc \cdot LH_2 - AMP + PPi$$

2. The intermediate complex Luc-D-luciferyl adenylate is oxidized by molecular oxygen with the formation of an excited enzyme-oxyluciferin-AMP complex and the release of carbon dioxide (CO_2):

$$Luc \cdot LH_2 - AMP + O_2 \rightarrow Luc \cdot AMP \cdot Oxyluciferin + CO_2$$

3. In the final step, energy loss from the excited complex produces photon emission and dissociation of the complex:

$$Luc \cdot AMP \cdot Oxyluciferin \rightarrow Luc + Oxyluciferin + AMP + h\nu$$

Photon emission is then recorded with a luminometer. A luminometer is made of photon-collecting apparatus (low-noise photomultiplier tube) and a computer station equipped with a software to allow data storage and analysis. The photomultiplier tube can be replaced by a high sensitive photo camera. Automated plate readers for luminescence (as well as fluorescence) measurement can also be used. Bioluminescence measurement in small animals is usually done with total body luminometers. Widespread use of bioluminescence techniques has made a tremendous impact in immunology and cancer where it has been used to investigate gene expression and track cancer cells in living animals. Luciferase reporter gene cloned downhill to many different promoters or fused in frame with the genes of interest allows to monitor the transcriptional activity of countless stimuli or to detect gene activation in many different pathophysiological conditions [2]. Alternatively, luciferase transfection has allowed investigation of intracellular ATP levels under many different metabolic conditions. Thanks to the possibility to fuse the luciferase gene in frame with leader sequences targeting various intracellular compartments, it has also become possible to monitor intra-organelle ATP changes [3]. Total body luminometry allows luminescence recording and analysis in those experimental settings where luciferase is used either as an in vivo intracellular reporter for cell tracking or as a probe to measure extracellular ATP. To perfect bioluminescence measurement of extracellular ATP, we have engineered a chimeric *firefly luciferase* selectively expressed on the outer aspect of the plasma membrane, and therefore named plasma membrane luciferase (pmeLUC) (Fig. 1) [4–6]. PmeLUC expression allows extracellular ATP monitoring in the extracellular compartment, notably in the close vicinity of the plasma membrane, thus pmeLUC-transfected cells can be used as in vivo reporters of the extracellular ATP concentration [7–9].

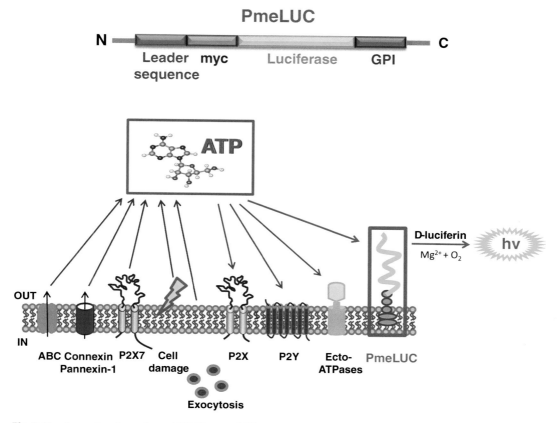

Fig. 1 Membrane topology of pmeLUC. The pmeLUC construct is made of the full-length coding sequence of luciferase (*yellow*) inserted in frame between the N-terminal leader sequence (*green*) and the C-terminal GPI anchor (*violet*) of the folate receptor. A c-myc tag (*light blue*) is added in frame for tracking purposes. The pmeLUC protein is targeted and localized to the extracellular aspect of the plasma membrane (from Falzoni et al., ref. 6)

2   Materials

2.1   Cells

Any cell type suitable to be transfected can be used as reporter of the extracellular ATP concentration with pmeLUC. We have used human HEK293 embryonic kidney cells, mouse CT26 colon carcinoma cells, mouse B16 melanoma cells, and many others. We will describe here the basic protocol for HEK293 cells.

2.2   Cell Culture

1. Cell media: Dulbecco's modified Eagle's medium (DMEM)-F12 medium containing 0.5 mM sodium pyruvate and 15 mM sodium bicarbonate, supplemented with 2 mM L-glutamine, 10 % heat-inactivated fetal bovine serum (FBS), 100 U/ml penicillin, and 100 μg/ml streptomycin.

2.3   Engineering of pmeLUC

1. pGL3 plasmid (Promega)
2. *Pst*I, *Not*I, *Xho*I, and *Xba*I restriction enzymes

3. pBSK+ vector (Stratagene)

4. *Pst*I fragment encoding the N-terminal leader sequence (26 aa) of the human folate receptor fused with a c-myc tag (10 amino acids)

5. *Pst*I fragment encoding the GPI anchor protein of the human folate receptor (28 amino acids)

6. pcDNA3 expression vector (Life Technologies)

7. *Leader-myc* forward primer with *Xba*I site (underlined):
1AF forward: 5′-GGT CTA GAG GAG AGC CAC CTC CT-3′

8. *Leader-myc* reverse primer without stop codon, with *Pst*I site (underlined):
1BR reverse: 5′-GGC TGC AGC AGG TCC TCC TCG CT-3′

9. Luciferase forward primer with *Pst*I site (underlined):
LucPstF forward: 5′-CCC TGC AGA TGG AAG ACG CCA AAA ACA TAA AGA AAG C-3′

10. Luciferase reverse primer with *Pst*I site (underlined) and without stop codon:
LucPstR reverse: 5′-GCT GCA GCC ACG GCG ATC TTT CCG CCC TTC TTG G-3′

11. 10× T4 ligase buffer: 200 mM Tris–HCl pH 7.6, 10 mM MgCl₂, 50 mM DTT, 500 μg/ml BSA, 10 mM ATP

12. T4 Ligase

13. Xl1-Blue *E. coli*

14. LB agar

15. Ampicillin

16. DNA extraction kit

2.4   Transfection

1. Complete cell culture medium (*see* Subheading 2.2).

2. HEK293 cells.

3. 2.5 M CaCl₂ solution in distilled water and filter sterilize. Store in 500 μl aliquots at –20 °C.

4. 2× HEPES-buffered saline (HBS-2X) solution: 280 mM NaCl, 50 mM HEPES base, 1.5 mM Na₂HPO₄, pH 7.12 with 0.5 N NaOH. Filter sterilize and store in 5 ml aliquots at –20 °C.

5. Phosphate buffered saline (PBS): 137 mM NaCl, 4.3 mM Na₂HPO₄ 7H₂O, 1.4 mM KH₂PO₄, 2.7 mM KCl, pH 7.4.

6. TRIS/EDTA (TE) buffer: 10 mM Tris–Cl pH 7.4, 1 mM EDTA pH 8.0 in filter-sterilized distilled water.

7. 10 cm Petri dishes for tissue culture.

8. 10 ml conical tubes.

9. 40 μg pmeLUC plasmid, resuspend in 450 μl of aqueous TE solution.

10. 40 μg of pcDNA3 empty vector, resuspend in 450 μl of TE solution.

11. G-418.

2.5 Immuno-fluorescence

1. 13 mm glass coverslips

2. 24-well cell culture plates

3. 0.01 % poly-L-lysine solution in water

4. PBS (*see* Subheading 2.4)

5. 2 % gelatin solution, type B, in distilled filter-sterilized H_2O

6. Anti c-myc-SC-40 monoclonal antibody (Santa Cruz Biotechnology)

7. Texas Red-conjugated goat anti-mouse IgG

8. FITC-conjugated anti-mouse antibody

9. Prolong Gold® antifade reagent (Life Technologies) or other suitable antifade reagents

10. 4 % paraformaldehyde solution in PBS

11. 10× solution Cell Fix™ (BD Biosciences)

2.6 In Vitro Extracellular ATP Measurement

1. 12-well cell culture plate.

2. DMEM-F12 medium.

3. 15 mg/ml D-luciferin stock solution in sterile PBS. Allow the luciferin solution to sit for a minimum of 15 min with gentle agitation prior to make 1 ml aliquots. Protect from light and store aliquots at −80 °C.

4. 100 mM adenosine-5′-triphosphate disodium salt (ATP) in 0.1 M Tris-base solution at pH 7.5. Store in 0.5 ml aliquots at −80 °C.

2.7 In Vivo Extracellular ATP Measurement

1. D-Luciferin.

2. Complete HEK293 cell culture medium (*see* Subheading 2.2).

3. RPMI 1640 supplemented with 2 mM L-glutamine, 1 mM Na-pyruvate, 10 % FBS, 100 U/ml penicillin, and 100 μg/ml streptomycin.

4. Isofluorane.

5. Syringe and 27-gauge needle.

6. 500 units apyrase from potato reconstitute in 1 ml of PBS buffer. Store in 50 μl aliquots at −80 °C.

7. PmeLUC-transfected HEK293 cells (HEK293-pmeLUC).

8. OVCAR-3 (human ovary carcinoma cell line) (*see* **Note 3**).

9. MZ2-MEL (human melanoma cell line) (*see* **Note 3**).

10. Athimic nude mice (*nude/nude* mice) 5–6 weeks old (Harlan Laboratories).

11. Total body luminometer (we use Perkin-Elmer Caliper IVIS 100 System™, but other compatible equipment is also suitable).

12. Apparatus for inhalatory anesthesia.

13. C57BL/6 or Balb/c 20 day old mice, weighing 12–14 g.

14. Dextran sulfate sodium salt powder from *Leuconostoc* spp. (MW > 500,000): dissolve 5 g of DSS in 100 ml of filtered drinking water, adjust the pH at 8.5. Store at 4 °C.

3 Methods

3.1 Engineering of pmeLUC

1. Firefly luciferase is amplified from the pGL3 plasmid (Promega) using the following primers: 5′-CCC TGC AGA TGG AAG CAA AAA ACA TAA AGA AAG G3′ (corresponding to the sequence encoding luciferase amino acids 1–9; *Pst*I site underlined) and 5′-GCT GCA GCC ACG GCG ATC TTT CCG CCC TTC TTG G3′ (corresponding to a 542–549 of luciferase cDNA without the stop codon; *Pst*I site underlined).

2. Thermal profile: denaturation at 95 °C for 30 min, hybridization at 62 °C for 30 min, polymerization at 72 °C for 2 min, repeat for 37 cycles.

3. Amplify the *leader-myc* sequence with primers described above using the following thermal profile: denaturation at 95 °C for 30 min, hybridization at 55 °C for 30 min, polymerization at 72 °C for 2 min, repeat for 32 cycles.

4. Transfer the PCR product to a pBSK+ vector, digest with *Pst*I, and insert in the right frame between a *Pst*I fragment encoding the N-terminal leader sequence of the human folate receptor (26 aa) fused with myc tag (10 aa) and a *Pst*I fragment of the GPI anchor protein (28 aa) to generate pmeLUC.

5. The whole final construct is excised by a *Not*I/*Xho*I or *Xba*I digestion and cloned into a pcDNA3 expression vector. For this purpose, a ligation reaction is performed with recombinant DNA and expression vector with a stoichiometric balance of 3:1 (Fig. 2).

6. The ligation buffer is made with 2 μl Buffer T4 ligase 10×, implemented with 1 μl T4 DNA ligation enzyme (2 U/μl) reconstituted in deionized water up to 20 μl to obtain the pmeLUC plasmid.

7. 100 ng of plasmid DNA is used to transform one aliquot of Xl1-Blue *E. coli* competent cells by thermic shock: 42 °C for 45 s followed by 2 min on ice.

CMV promoter: bases 209-863
T7 promoter: bases 864-882
Polylinker: bases 889-994
Sp6 promoter: bases 999-1016
BGH poly A: bases 1018-1249
SV40 promoter: bases 1790-2115
SV40 origin of replication: bases 1984-2069
Neomycin ORF: bases 2151-2945
SV40 poly A: bases 3000-3372
ColE1 origin: bases 3632-4305
Ampicillin ORF: bases 4450-5310

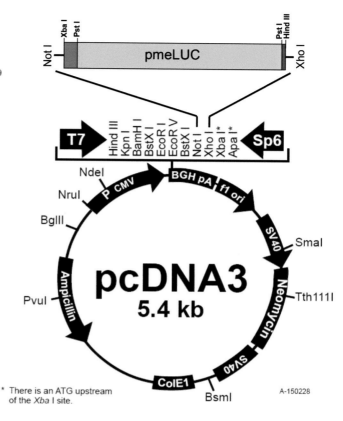

Fig. 2 Map of the pmeLUC plasmid

8. Thereafter, the bacterial cell suspension is cultured overnight at 37 °C in LB agar in the presence of 100 µg/ml ampicillin.

9. The day after, the bacterial cell suspension is precipitated by centrifugation at $3000 \times g$ for 20 min.

10. DNA is extracted with DNA extraction kit from Qiagen as per manufacturer instruction.

3.2 Cell Transfection

1. Plate HEK293 (3×10^6 cells/plate) into 10-cm tissue culture Petri dish 24 h before transfection. Rinse and add fresh DMEM F12 medium 4 h before transfection.

2. Prepare tube 1 solution (DNA-Ca) as follows: add 50 µl of $CaCl_2$ solution to 40 µg pmeLUC-pcDNA3 plasmid resuspended in 450 µl of aqueous (TE solution or DNAse/RNAse-free H_2O).

3. Prepare tube 2 solution: 500 µl HBS-2X solution.

4. Add the DNA-Ca solution (tube 1) dropwise with a Pasteur pipet to tube 2 and vortex immediately for a few seconds. Allow the precipitate to sit 30 min at room temperature.

5. Add the precipitate to the HEK293 cell culture plate, mix gently, and incubate overnight.

6. The day after remove medium, wash cells twice with 10 ml of warm (37 °C) PBS, and add 10 ml of complete DMEM-F12. For stable transfection, allow HEK293 cells to double twice before plating in selection medium.

3.3 Selection of Stably Transfected Cell Clones

1. For stable transfection, the cell culture is supplemented with 0.8 mg/ml of the antibiotic G 418 three days after transfection and kept in its continuous presence for 2 weeks.

2. Thereafter, positive clones are obtained by limiting dilution: 100 cells are resuspended in 10 ml of 0.2 mg/ml G418-supplemented DMEM-F12 medium; 100 μl of this cell suspension are added to each well of a 96-well cell plate. Place the cell plate in a 5 % CO_2, humidified incubator at 37 °C.

3. Check clones with a phase contrast microscope at 20× magnification after 5–6 days. Check each well and mark those well that contain only one colony.

4. Transfer colonies to a T-25 culture flask in 0.2 mg/ml G418-supplemented DMEM-F12 medium and wait for colony growth.

3.4 In Vitro Validation of Transfected Cells by Immuno-fluorescence

1. 13 mm glass coverslips are placed into single wells of a 24 well culture plate.

2. Add 200 μl of poly-L-lysine to each well for coverslip coating and incubate 1 h at room temperature.

3. Rinse coverslips twice with sterile H_2O. Allow coverslips to dry completely and sterilize under UV light for at least 4 h.

4. Plate 2×10^4 HEK293 pmeLUC cells onto each coverslip. Allow cells to adhere and fix with formaldehyde, 4 % in PBS, for 30 min.

5. Rinse coverslips three times with PBS.

6. Incubate coverslips for 30 min with 0.2 % gelatin in PBS to block nonspecific binding sites.

7. Immunostaining is carried out for 1 h at 37 °C with a commercial monoclonal Sc-40 antibody against the c-myc epitope tag diluted at 1:100 in 0.2 % gelatin-supplemented PBS.

8. Discharge the antibody solution, add fresh PBS, and incubate for 5 min. Repeat washing three times, for a total of 15 min.

9. Immunodetection is carried out using Texas Red-conjugated goat anti-mouse IgG diluted 1:50 in PBS supplemented with 0.2 % gelatin. Leave the coverslips in this solution for 1 h at room temperature in the dark.

10. Discharge the secondary antibody, add fresh PBS, and incubate for 5 min. Repeat washing three times, for a total of 15 min.

11. After immunostaining, cells are mounted with a drop of Prolong Gold antifade reagent and analyzed with a fluorescence microscope at 60× or 100× magnifications, e.g., a Zeiss LSM 510 confocal microscope.

12. HEK293 cells, mock transfected with the empty pcDNA3 vector are used as a negative control.

3.5   In Vitro Validation of Transfected Cells by FACS Analysis

1. Resuspend 1×10^6 HEK293-pmeLUC cells in 1 ml of PBS.

2. Incubate HEK293-pmeLUC cells for 1 h at 4 °C with monoclonal Sc-40 antibody against the c-myc epitope tag diluted at 1:100.

3. Rinse two times with cold PBS.

4. Incubate cells at 4 °C for 1 h with a FITC-conjugated antimouse antibody at a 1:50 dilution in PBS. Protect from light.

5. Rinse two times with cold PBS.

6. Fix cells with 100 μl of 10× Cell Fix™ diluted in 0.9 ml of cold PBS.

7. Acquire fluorescence with a flow cytometer (e.g., Becton Dickinson BD FACSscan), and analyze data with BD Cell Quest software. HEK293 cells, mock- transfected with the empty pcDNA3vector, are used as a negative control (*see* **Note 1**).

3.6   Analysis of pmeLUC Expression and Function by In Vitro Luminescence Recording with a Total Body Luminometer

1. Seed 7×10^4 HEK293-pmeLUC cells into 12-well plates (Becton Dickinson Biosciences) in DMEM-F12 medium. Allow to adhere overnight in a 5 % CO_2, 37 °C, humidified incubator.

2. The following day, add D-luciferin sodium salt (*see* **item 3**, Subheading 2.6) to each well (8 μl, 120 μg/well), and, 3 min later, start luminescence acquisition in a luminometer (e.g., IVIS 100). Set acquisition time and binning at 1 min/plate and 4, respectively.

3. Build a calibration curve by consecutive additions of increasing (from 0.001 to 1 mM) ATP concentrations to one or more wells. Start a new acquisition after each ATP addition.

4. As a control, add 5 U apyrase to one or more wells and start acquisition. Apyrase addition should cause a large luminescence decrease (*see* **Note 2**).

5. A region of interest (ROI) is manually selected on each well. Keep the area of ROI constant and record the intensity as photons per seconds.

3.7 Analysis of pmeLUC Expression and Function by In Vitro Luminescence Recording with a Plate Reader Luminometer

1. Plate the HEK293-pmeLUC cells as above.

2. Add 8 μl of D-luciferin solution (*see* **item 3**, Subheading 2.6) to each well.

3. Add increasing ATP concentrations solution to consecutive wells (calibration).

4. Place the plate in the plate reader luminometer and start the acquisition. Set the counting time at 7 s/well. Triplicates are recommended.

5. Record luminescence in counts per second (CPS) and express as function of ATP concentration.

3.8 In Vivo Analysis of Extracellular ATP

PmeLUC-transfected reporter cells can be used to monitor the extracellular ATP concentration in several experimental settings. In healthy mice, the extracellular ATP concentration is negligible, i.e., in the low nanomolar range, and therefore below pmeLUC detection limit. Thus, best examples of extracellular ATP measurements by pmeLUC are given in experimental models of inflammation and cancer.

3.9 Measurement of Extracellular ATP Levels in a Model of Experimental Colitis Induced with Dextran Sulfate

1. Inject mice (C57Bl/6 or Balb/c) i.p. with 2×10^6 HEK293-pmeLUC cells in 200 μl of DMEM-F12 the day before starting DSS administration to record basal peritoneal ATP levels.

2. Gently massage the abdomen to allow pmeLUC cell distribution throughout the peritoneum.

3. Fifteen minutes later, inject each mouse i.p. with 150 mg/kg D-luciferin (3 mg/mouse) of D-luciferin-containing PBS in a final volume of 200 μl. Wait 15 min to allow luciferin distribution throughout the mouse tissue.

4. Anesthetize mice with isofluorane (2 % in 1 L/min oxygen).

5. Place the animals (abdominal view) into a total body luminometer (e.g., IVIS 100 System™). Set acquisition time and binning at 3 min/acquisition and 4, respectively. Up to three mice each time can be acquired at the same time.

6. For quantification select manually a region of interest (ROI) around peritoneum. Record ROI luminescence intensity as photons per seconds.

7. Analyze data with the Living Image software.

8. The day after, supplement mice drinking water with 5 % DSS (dextran sulfate solution) for at least 7 days.

9. Feed control mice with regular, DSS-free, drinking water.

10. Replace drinking water, with or without DSS, daily.

11. Check mice wellness daily: measure body weight and inspect stools to monitor rectal bleeding (indications of colitis).

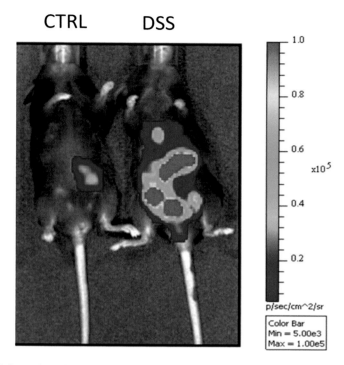

Fig. 3 Detection of experimental colitis by HEK293-pmeLUC cell inoculated i.p. in DSS-treated C57Bl/6 mouse. Mouse on the right received DSS and HEK293-pmeLUC cells. Mouse on the *left* (control) received only HEK293-pmeLUC cells

12. Repeat **steps 1–7** every 2 days for at least 10 days.

13. Colitis shows as an increased luminescence emission of the colon (Fig. 3).

3.10 Measurement of the ATP Content of the Tumor Microenvironment in a Model of Peritoneal OVCAR-3 Ovarian Carcinoma

1. Resuspend 1.5×10^6 OVCAR-3 cells in 200 μl of RPMI.

2. Inject i.p. into *nude/nude* mice the OVCAR-3 cell suspension (200 μl in each mouse) with a syringe fitted with a 27-gauge needle. A minimum of five mice is recommended.

3. Wait 20 days for tumor growth.

4. Inject i.p. in each mouse 2×10^6 HEK293-pmeLUC cells in 200 μl of DMEM-F12.

5. Twenty-four hours later, i.p. inject each mouse with 200 μl of D-luciferin-containing PBS (150 mg/kg D-Luciferin, 3 mg/mouse). Wait 15 min and anesthetize the mice with isofluorane (2 % in 1 L/min oxygen).

6. Place the animals (abdominal view) under continuous anesthesia into the IVIS 100™ Luminometer. Set acquisition time and binning at 3 min/acquisition and 4, respectively. Up to three mice can be acquired at the same time.

7. Repeat **steps 5** and **6** every 2 days for at least 16 days.

8. For quantification, select manually a region of interest (ROI) including the tumor area, and record ROI luminescence intensity as photons per seconds.

9. Analyze data with Living Image software (*see* **Note 4**).

10. Inoculate HEK293-pmeLUC cells into healthy *nude/nude* mice as a control (*see* **Note 5**).

3.11 Measurement of the ATP Content of the Tumor Microenvironment in a Model of MZ2-MEL Melanoma

1. Resuspend 8×10^6 MZ2-MEL cells in 200 µl of RPMI.

2. Inject into *nude/nude* mice the MZ2-MEL suspension i.p. (200 µl in each mouse) with a syringe fitted with a 27-gauge needle. A minimum of five mice is recommended.

3. Inject the cell suspension subcutaneously (s.c.) in the right dorsal hip of each mouse with a syringe fitted with a 27-gauge needle.

4. Wait 10–15 days for tumor growth, or until tumor mass has reached a size of 1.5×1.5 cm.

5. Inject 2×10^6 HEK293-pmeLUC cells in 200 µl of DMEM-F12 into the tumor mass.

6. As a control, inject 2×10^6 HEK293-pmeLUC cells into the contralateral (left) dorsal hip (healthy tissue) of each mouse.

7. Inject each mouse i.p. with 150 mg/kg D-luciferin (3 mg/ mouse).

8. Fifteen minutes after, anesthetize the mice with isofluorane (2 % in 1 L/min oxygen).

9. Place the animals (dorsal view) into the luminometer. Set acquisition time and binning at 3 min/acquisition and 4, respectively.

10. Up to three mice at the same time can be acquired.

11. Repeat **steps 6** and **7** every 2 days for at least 10 days.

12. For quantification, select manually a region of interest (ROI) including the tumor area. Draw ROI with the same size area also on the left dorsal hip (healthy tissue is used as a negative control). Record ROI luminescence intensity as photons per seconds (Fig. 4).

13. Analyze data with Living Image software.

3.12 HEK293-pmeLUC Cell Validation with Apyrase

1. Induce a melanoma tumor as describe above (*see* Subheading 3.11).

2. When tumor mass has reached a size of 1.5×1.5 cm, inject 2×10^6 HEK293 pmeLUC cells (resuspended in 200 µl of DMEM-F12) into the tumor mass.

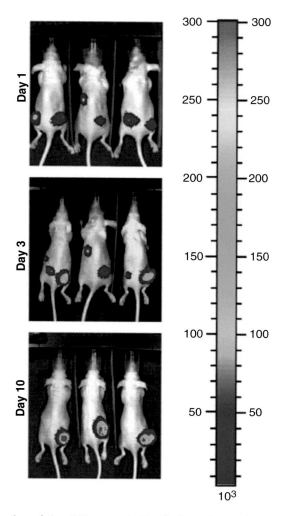

Fig. 4 Detection of the ATP concentration in the microenvironment of MZ2-MEL tumor intra-mass injected with HEK293-pmeLUC cells (from Pellegatti et al., ref. 5)

3. Four days later, inject into the tumor mass 20 U of apyrase dissolved in 100 μl of sterile PBS. Inject sterile PBS alone (vehicle) in control tumor-bearing mice.

4. Acquire and measure luminescence as described in Subheading 3.11.

4   Notes

1. To enhance plasma membrane expression of the pmeLUC construct, cells can be overnight incubated in the presence of 1 mM DTT, or kept at room temperature (21 °C) for 2 h

before immunostaining or luminescence recording. These treatments do not affect luciferase activity, and maximize pme-LUC surface expression by enhancing transport to the plasma membrane and slowing recycling.

2. Apyrase, a cell-impermeant ATP-hydrolyzing enzyme, is used as a control of the extracellular location of pmeLUC probe.

3. OVCAR-3 and MZ2-MEL cell lines are cultured in RPMI 1640 with L-glutamine, Na-pyruvate, and $NaHCO_3$ (Sigma Aldrich) supplemented with 10 % heat-inactivated FBS, 100 U/ml penicillin, and 100 μg/ml streptomycin. Cells are cultured in Falcon T75 cell culture flasks and split once they reach 80–90 % confluence by treatment with 0.025 % Tripsin and 0.025 % EDTA in PBS buffer.

4. In tumor-bearing mice luminescence is very bright shortly after pmeLUC inoculation, because of the ATP-rich inflammatory environment within the peritoneal cavity. Then the signal progressively attenuates and localizes to discrete abdominal foci (metastases). Postmortem analysis and luminometry reveal numerous light-emitting metastasis on the abdominal wall. Moreover, histologic analysis of metastases with anti-luciferase Abs shows that HEK293 pmeLUC cells infiltrate tumor masses and report the ATP concentration in the tumor microenvironment.

5. Control mice: Healthy nude mice were injected i.p. with 2×10^6 HEK293 pmeLUC cells and monitored for 3 months. No significant luminescence emission can be detected over this extended time span. This underlines that in healthy tissue the ATP concentration in the extracellular environment is very low, certainly below the threshold for detection by HEK293 pmeLUC (1–5 μM).

References

1. Wilson T, Hastings JW (1998) Bioluminescence. Annu Rev Cell Dev Biol 14:197–230

2. Contag CH, Bachmann MH (2002) Advances in in vivo bioluminescence imaging of gene expression. Annu Rev Biomed Eng 4: 235–260

3. Jouaville LS, Pinton P, Bastianutto C, Rutter GA, Rizzuto R (1999) Regulation of mitochondrial ATP synthesis by calcium: evidence for a long-term metabolic priming. Proc Natl Acad Sci U S A 96:13807–13812

4. Pellegatti P, Falzoni S, Pinton P, Rizzuto R, Di Virgilio F (2005) A novel recombinant plasma membrane-targeted luciferase reveals a new pathway for ATP secretion. Mol Biol Cell 16:3659–3665

5. Pellegatti P, Raffaghello L, Bianchi G, Piccardi F, Pistoia V, Di Virgilio F (2008) Increased level of extracellular ATP at tumor sites: in vivo imaging with plasma membrane luciferase. PLoS One 3:e2599

6. Falzoni S, Donvito G, Di Virgilio F (2013) Detecting adenosine triphosphate in the pericellular space. Interface Focus 3:20120101

7. Wilhelm K, Ganesan J, Muller T, Durr C, Grimm M, Beilhack A et al (2010) Graft-

versus-host disease is enhanced by extracellular ATP activating P2X7R. Nat Med 16: 1434–1438

8. Bianchi G, Vuerich M, Pellegatti P, Marimpietri D, Emionite L, Marigo I et al (2014) ATP/P2X7 axis modulates myeloid-derived suppres-sor cell functions in neuroblastoma microenvi-ronment. Cell Death Dis 5, e1135

9. Loo JM, Scherl A, Nguyen A, Man FY, Weinberg E, Zeng Z et al (2015) Extracellular metabolic energetics can promote cancer pro-gression. Cell 160:393–406

Chapter 8

Measuring NLR Oligomerization I: Size Exclusion Chromatography, Co-immunoprecipitation, and Cross-Linking

Sonal Khare, Alexander D. Radian, Andrea Dorfleutner, and Christian Stehlik

Abstract

Oligomerization of nod-like receptors (NLRs) can be detected by several biochemical techniques dependent on the stringency of protein–protein interactions. Some of these biochemical methods can be combined with functional assays, such as caspase-1 activity assay. Size exclusion chromatography (SEC) allows separation of native protein lysates into different sized complexes by fast protein liquid chromatography (FPLC) for follow-up analysis. Using co-immunoprecipitation (co-IP), combined with SEC or on its own, enables subsequent antibody-based purification of NLR complexes and associated proteins, which can then be analyzed by immunoblot and/or subjected to functional caspase-1 activity assay. Chemical cross-linking covalently joins two or more molecules, thus capturing the oligomeric state with high sensitivity and stability. Apoptosis-associated speck-like protein containing a caspase activation domain (ASC) oligomerization has been successfully used as readout for NLR or AIM2-like receptor (ALR) inflammasome activation in response to various pathogen- or damage-associated molecular patterns (PAMPs or DAMPs) in human and mouse macrophages and THP-1 cells. Here, we provide a detailed description of the methods used for NLRP7 oligomerization in response to infection with *Staphylococcus aureus* (*S. aureus*) in primary human macrophages, co-immunoprecipitation and immunoblot analysis of NLRP7 and NLRP3 inflammasome complexes, as well as caspase-1 activity assays. Also, ASC oligomerization is shown in response to dsDNA, LPS/ATP, and LPS/nigericin in mouse bone marrow-derived macrophages (BMDMs) and/or THP-1 cells or human primary macrophages.

Key words Cross-linking, Size exclusion chromatography, Co-immunoprecipitation, Protein–protein interaction, Oligomerization, Caspase-1 activity assay, Nod-like receptor, NLR, Inflammasome

---

1 Introduction

Nod-like receptors (NLRs) assemble into large oligomeric signaling complexes, such as inflammasomes and nodosomes [1, 2]. Several biochemical methods, which are based on the detection of protein–protein interactions, are established to detect and quantify the conversion of monomeric proteins into active oligomeric protein complexes [3–5].

Francesco Di Virgilio and Pablo Pelegrín (eds.), *NLR Proteins: Methods and Protocols*, Methods in Molecular Biology, vol. 1417, DOI 10.1007/978-1-4939-3566-6_8, © Springer Science+Business Media New York 2016

Size exclusion chromatography (SEC) is based on the separation of native protein complexes according to their size by migration through a gel matrix, which consists of spherical beads containing pores of a specific size distribution [6]. When molecules of different sizes are included or excluded within the matrix, it results in separation of these molecules depending on their overall sizes. Small molecules diffuse into pores and are retarded depending on their size, whereas large molecules, which do not enter the pores, are eluted with the void volume of the column. As the molecules pass through the column, they are separated on the basis of their size and are eluted in the order of decreasing molecular weights. Here we describe SEC of the oligomerized NLRP7 complex in response to *S. aureus* infection of human primary macrophages [7].

Immunoprecipitation (IP) and co-immunoprecipitation (co-IP) are routinely used techniques to study protein–protein interactions and to identify novel members of protein complexes [3, 8, 9]. Both techniques use an immobilized antibody specific to the antigen/protein of interest. While IP is designed to purify a single antigen, co-IP is suited to isolate the specific antigen/protein as well as to co-purify any other associated proteins, which are then separated by SDS/PAGE and detected by immunoblotting. Interacting proteins might include complex partners, cofactors, signaling molecules, etc. The strength of the interaction between proteins may range from highly transient to very stable interactions. While studying these interactions by co-IP, there are a number of factors which should be taken under consideration, e.g., specificity of the antibody, optimization of the binding and wash conditions, posttranslational modifications, etc. Here, we describe a co-IP protocol for the endogenous ASC–NLRP3 complex from THP-1 cells and BMDMs and the ASC–NLRP7 complex from human primary macrophages, as the recruitment of ASC to these NLRs is a readout for inflammasome assembly. A particularly useful approach is the combination of SEC with co-IP to allow the analysis of complexes within a specific size fraction, for example, for analyzing NLR-containing complexes within high molecular weight fractions.

This analysis further enables the detection of caspase-1 within inflammasomes and allows quantification of its activity, when combined with caspase-1 activity assays. Caspase-1, also known as interleukin (IL)-1β-converting enzyme (ICE), is a cysteine protease and is the downstream effector molecule that becomes activated within inflammasomes subsequent to the activation of several NLRs [10]. The active 20 kDa and 10 kDa hetero-tetrameric caspase-1 is derived from the auto-proteolytically cleaved 45 kDa pro-enzyme (zymogen) [11, 12]. Subsequently, the caspase-1 substrate pro-IL-1β (35 kDa) is converted into the biologically active form (17 kDa) [13–15]. Here we describe two assays that determine caspase-1 activity, which are routinely used in our laboratory [7]. First, a sensitive fluorometric assay that quantifies caspase-1 activity within the NLRP7 inflammasome, where the preferential recognition of the tetrapeptide sequence YVAD by caspase-1 is utilized in combination with the detection of the fluorescent substrate 7-amino-4-trifluoromethyl coumarin (AFC) [16]. YVAD-AFC emits blue light (400 nm), but once the substrate is cleaved by caspase-1, the free AFC emits yellow-green fluorescence (505 nm), which can be quantified in a plate reader with fluorescence capabilities and the appropriate filter sets. Second, the caspase-1 substrate pro-IL-1β is converted into mature IL-1β, which can be detected by Western blot analysis [7].

Chemical cross-linking covalently joins two or more molecules [17]. Cross-linking reagents (or cross-linkers) consist of two or more reactive ends. This enables cross-linkers to chemically attach to specific functional groups (e.g., sulfhydryls, primary amines, carboxyls, etc.) on proteins or other molecules. Cross-linker-mediated attachment between groups on two different protein molecules leads to intermolecular cross-linking. This cross-linking results in the stabilization of protein–protein interactions. Cross-linkers can be selected on the basis of their chemical reactivity and properties, like chemical specificity, water solubility, membrane permeability, etc. [17]. Here, we describe the cross-linking of nucleated and polymerized ASC molecules using the membrane-permeable, nonreversible cross-linker disuccinimidyl suberate (DSS), which contains an amine-reactive N-hydroxysuccinimide (NHS) ester at each end of an 8-carbon spacer arm (see **Note 1**). ASC oligomerization has been successfully used as a readout for NLR or ALR inflammasome activation in response to various PAMPs and DAMPs in primary macrophages as well as monocytes [18, 19]. We describe ASC oligomerization in response to poly(dA:dT) (dsDNA) as well as LPS/ATP or LPS/nigericin in mouse BMDM and THP-1 cells [20].

2   Materials

All solutions should be prepared using ultrapure water (sensitivity of 18 MΩ·cm at 25 °C) and analytical grade reagents.

2.1   Cell Culture and Antibodies

1. Appropriate culture medium for THP-1 cells, primary human macrophages, BMDMs, and HEK293 cells.

2. Lipofectamine 2000 (Invitrogen) or transfection reagent of choice.

3. Phosphate-buffered saline (PBS) solution.

4. 100/60 mm tissue culture dishes.

5. 1.5 and 2 ml microcentrifuge tubes.

6. 15 ml centrifuge tubes.

7. Cell scrappers.

8. Refrigerated tabletop centrifuge with 1.5 ml microcentrifuge tube rotor.

9. Tabletop centrifuge with 15 ml tube rotor.

10. Heat block.

11. 10× protease inhibitor cocktail (we use the cocktail from Roche, but other cocktails could be also used).

12. 0.1 M phenylmethylsulfonyl fluoride (PMSF) in ethanol.

13. Poly(dA:dT).

14. Ultrapure *E. coli* lipopolysaccharide (LPS), serotype 0111:B4.

15. ATP.

16. Nigericin.

17. Laemmli sample loading buffer: 60 mM Tris–HCl, pH 6.8, 2 % sodium dodecyl sulfate (SDS), 100 mM dithiothreitol (DTT), 10 % glycerol, and 0.01 % bromophenol blue.

18. SDS/PAGE and blotting equipment and materials.

19. Anti-ASC antibody (we use Santa Cruz, sc-22514-R).

20. Anti-NLRP3 antibody (we use Adipogen, Cryo-2, AG-20B-0014).

21. Anti-NLRP7 antibody (we use Imgenex, IMG-6357A).

2.2 Size Exclusion Chromatography (SEC)

1. Fast protein liquid chromatography (FPLC) equipment (we use a Bio-Rad BioLogic LP Chromatography System, but other systems are also suitable).

2. Gel filtration column. We use GE Healthcare HiPrep 16/60 Sephacryl S-300 High-Resolution Column (matrix, 50 μm allyl dextran and N,N'-methylenebisacrylamide; bed dimension, 16×600 mm; bed volume, 120 ml). Other similar columns are also suitable.

3. System tubing of 1.6 mm internal diameter.

4. Fraction collector.

5. 5 ml round bottom tubes.

6. Separation buffer: 50 mM Tris, pH 7.4, 150 mM NaCl.

7. Lysis buffer: 10 mM $Na_4P_2O_7$, 10 mM NaF, 20 mM Tris, pH 7.4, 150 mM NaCl, 1 % octylglucoside, 1 mM PMSF, and 1× protease inhibitor cocktail.

8. Trichloroacetic acid (TCA).

9. Acetone.

10. 20 % ethanol.

11. Gel filtration standard (we use Bio-Rad 151-1901).

12. Dounce homogenizer.

13. Sonicator.

14. 3 ml syringe using an 18½-gauge needle.

15. 0.45 µM syringe filter.

2.3   Co-immunoprecipitation (co-IP)

1. IP lysis buffer: 50 mM HEPES, pH 7.4, 150 mM NaCl, 10 % glycerol, 2 mM EDTA, 0.5 % Triton X-100, 1× protease inhibitor cocktail.

2. Specific antibody against NLR of interest (*see* Subheading 2.1).

3. Protein A/G-Sepharose (we use the Protein A/G-Sepharose® 4B from Life Technologies).

2.4   Fluorometric Caspase-1 Activation Assay

1. 10 mM Ac-YVAD-AFC. This is a stock solution and should be prepared in dimethyl sulfoxide (DMSO); protect solution from light.

2. Caspase-1 lysis buffer: 150 mM NaCl, 20 mM Tris, pH 7.5, 0.2 % Triton X-100; **add freshly**: 1 mM DTT.

3. Caspase assay buffer: 50 mM HEPES, 7.2–7.4, 50 mM NaCl, 1 mM EDTA, 0.1 % CHAPS, 10 % sucrose; **add freshly**: 10 mM DTT, 100 µM substrate Ac-YVAD-AFC.

2.5   In Vitro Caspase-1 Activation Assay

1. pro-IL-1β cDNA.

2. Anti-IL-1β antibody suitable for Western blotting (e.g., we use cell signaling antibody clone 3A6, 12242).

2.6   Cross-Linking

1. 100 mM disuccinimidyl suberate (DSS). This is a stock solution and should be prepared in DMSO (**prepare freshly just before adding to the lysates**).

2. Lysis buffer: 20 mM HEPES–KOH, pH 7.5, 150 mM KCL, 1 % NP-40, 0.1 mM PMSF, 1× of protease inhibitor cocktail, 1 mM sodium orthovanadate (PMSF should be added to the lysis buffer just before cell lysis).

3   Methods

3.1   Cell Culture and Activation

1. Culture THP-1 cells, primary human macrophages, or BMDM in the appropriate culture dish and with your established media.

2. To activate the AIM2 inflammasome, transfect cells with poly(dA:dT) using Lipofectamine 2000 (follow the manufacturer's protocol for the transfection) for 5 h (for cross-linking, *see* **Note 2**). To activate the NLRP3 inflammasome with LPS/ATP, treat cells with 100 ng/ml of LPS for 4 h followed by a pulse with 5 mM ATP for 20 min. or incubation with nigericin (5 µM)

for 45 min. Infect cells with *S. aureus* (MOI = 3) or *Listeria monocytogenes* (MOI = 12) at 37 °C for 45 min to activate multiple inflammasomes, including NLRP3, NLRP7, and AIM2 [7, 20]. Extracellular bacteria will be eliminated with gentamicin (50 mg/ml) for a total time of 90 min [7] (*see* **Note 3**).

3.2 Size Exclusion Chromatography (SEC)

1. Connect the column according to the manufacturer's instructions and verify proper bubble-free packing of the column.

2. Equilibrate the column with ½ column volume (CV) ddH$_2$O (60 ml) at a flow rate of 0.5 ml/min, followed by 2 CV (240 ml) separation buffer at 1 ml/min at 4 °C.

3. Activate 5×10^7 cells in 60 mm culture dishes containing 5×10^6 macrophages per plate for 90 min (*see* **Note 4**).

4. Aspirate the culture medium and wash cells twice with ice-cold 1× PBS.

5. Lyse cells in 200 µl lysis buffer per plate, pool lysate from all plates into a 15 ml conical tube, and incubate for 20 min on ice.

6. Prior to homogenization, prechill a glass of 2 ml Dounce tissue grinder on ice, transfer lysate to the cooled tissue grinder, and homogenize through 30 strong strokes, carefully avoiding bubble formation.

7. Following homogenization, place the lysate in a 2 ml microcentrifuge tube and spin at $12,000 \times g$ for 30 min to clarify the lysate.

8. Transfer the cleared lysates to a fresh tube and then draw it into a 3 ml syringe using an 18½-gauge needle. Perform a second clarification step to remove any remaining debris using a 0.45 µM syringe filter. Draw the final lysate in a new 3 ml syringe for injection into the chromatography system or store it at 4 °C until injection.

9. Carefully examine the lysate within the syringe for any air bubbles prior to injection, and if necessary remove it by gently flicking the syringe.

10. Insert the syringe into the injection valve in the valve controller and inject the lysate into a sample loop, which needs to fit the entire lysate (we usually use a 2.5 ml sample loop) prior to introducing it into the separation column. During injection, the flow of the chromatography system bypasses the sample loop. Adjust the flow rate to 0.5 ml/min, and use the valve selection device to divert the flow into the sample loop, which is subsequently connected to the separation column and the rest of the system. Take care to avoid any introduction of air bubbles into the system.

11. Start an automated collection program:

(a) 0.5 ml/min flow rate for 72 min, diverting flow to disposal (column dead volume).

(b) 0.5 ml/min flow rate for 216 min, diverting flow to fraction collector; collect 9 min/fraction (4.5 ml) for a total of 24 fractions.

(c) Divert flow to waste for another 200 min at 0.5 ml/min to remove any small molecules remaining in the column.

12. Continuously monitor fractions by UV absorbance.

13. Initiate any subsequent injections at this point, repeating **steps 10–12**.

14. At this step, either continue with co-IP (*see* Subheading 3.3 below) or continue for TCA precipitation and Western blot analysis.

15. Divide each fraction into three 1.5 ml aliquots in 2 ml microcentrifuge tubes.

16. TCA precipitates proteins by adding 500 μl 100 % w/v trichloroacetic acid per tube. Mix tubes and incubate on ice for at least 10 min and collect proteins by centrifugation at $14,000 \times g$ for 30 min. White protein precipitates are present at the bottom of the tubes. Discard the supernatants and wash pellets with 0.5 ml of ice-cold acetone with vortexing and pool the contents of three tubes corresponding to the same fraction and spin at $14,000 \times g$ for 10 min.

17. Aspirate the acetone supernatant and wash the pellets twice with 1 ml of ice-cold acetone as above.

18. Aspirate the acetone and keep the tubes uncapped at 30 °C in a heat block for 10 min to evaporate any remaining acetone (*see* **Note 5**).

19. Resuspend protein pellets in 50 μl 1.5× Laemmli buffer (*see* **Note 6**). Vigorously vortex the samples to fully resuspend any insoluble portion of the pellet. Sonicate samples in a water bath to ensure maximum dissolution of the protein pellet. Boil samples in a 95 °C heat block for 10 min before separating by SDS/PAGE (a 10 % acrylamide gels is usually well suited) and analysis by immunoblotting (*see* **Note 7**).

20. To match fractions to a particular molecular weight, either pre-run or post-run a gel filtration molecular weight standard under the same conditions. The protein standards can be detected by UV absorbance and matched to a particular fraction. We use a lyophilized protein mixture from Bio-Rad consisting of thyroglobulin, bovine γ-globulin, chicken ovalbumin, equine myoglobin, and vitamin B12 (MW range from 1350 to 670,000 Da, pI 4.5–6.9), which we dissolve in lysis buffer.

21. Regenerate the column after each run with one CV (120 ml) separation buffer (at 1 ml/min at 4 °C) and four CV (480 ml) ddH$_2$O, followed by four CV (480 ml) 20 % ethanol, and store in 20 % ethanol at 4 °C in an upright position.

3.3 Co-immunoprecipitation (co-IP)

1. Plate 1×10^7 cells in a 100 mm tissue culture dish (*see* **Note 8**).

2. Wash the adherent cells (primary human macrophages and BMDM) with ice-cold PBS. Add 1 ml of ice-cold lysis buffer. Keep the plate on ice and make sure the lysis buffer is distributed evenly. In case of THP-1 cells (suspension cells), transfer the cells to a 15 ml tube and centrifuge at $400 \times g$ for 10 min at RT. Add 1 ml of ice-cold lysis buffer and transfer the cell suspension to a 1.5 ml tube. Incubate the cells in lysis buffer for 30 min on ice.

3. Centrifuge the cell lysates at $10,000 \times g$ for 15 min at 4 °C.

4. Transfer the cleared lysates to a fresh 1.5 ml microcentrifuge tube. Alternatively, use a particular fraction or pooled fractions from SEC as input for the co-IP (from **step 14** of Subheading 3.2).

5. Preclear lysates with 1 μg control IgG and 5 μl of Protein A/G-Sepharose beads.

6. Centrifuge the samples at $2500 \times g$ at 4 °C for 2 min.

7. Transfer the supernatant to a fresh 1.5 ml microcentrifuge tube.

8. Add 1 μg of specific antibody to the NLR of interest or ASC to the cleared lysates and incubate at 4 °C with rotation for 1 h (*see* **Note 9**).

9. Add 5 μl of Protein A/G-Sepharose beads to the lysate–antibody mix and incubate at 4 °C with rotation for at least 2 h or overnight.

10. Centrifuge the samples at $2500 \times g$ at 4 °C for 2 min. Remove the supernatant and wash the beads three times with 1 ml lysis buffer (*see* **Note 10**).

11. Extract the immunoprecipitated proteins by adding 60 μl of Laemmli sample loading buffer and incubating at 95 °C for 10 min.

12. Centrifuge the samples at $2500 \times g$ for 2 min to pellet the Sepharose beads.

13. Run the cleared samples on the SDS/PAGE gel and detect the immunoprecipitated proteins and the potential interacting partners by immunoblotting using specific antibodies.

3.4 co-IP of Overexpressing NLR Proteins with ASC

This is an alternative approach which has been previously used to analyze ASC–NLRP3 and ASC–NLRP7 interactions [7, 20, 21].

1. Transfect a 100 mm tissue culture dish of HEK293 cells with myc-ASC, control GFP, GFP-NLRP3, or GFP-NLRP7, adjusted to yield comparable expression (*see* **Note 11**).

2. 36 h post-transfection, lyse the cells in IP lysis buffer.

3. Proceed for the co-IP using protein-/tag-specific antibody as described in Subheading 3.3 (*see* **Note 12**).

3.5 Fluorometric Caspase-1 Activation Assay

1. Pellet from 1 to 5×10^6 cells at $400 \times g$ for 10 min in a centrifuge at 4 °C (*see* **Note 13**).

2. Wash the cells once with ice-cold PBS.

3. Add caspase lysis buffer with fresh DTT (use the same volume as the cell pellet, as a very concentrated cell lysate is required: about 50 μl), resuspend and keep on ice for 10 min.

4. Spin for 5 min in a refrigerated centrifuge at $14,000 \times g$ and 4 °C (full speed) and transfer supernatant into a fresh, prechilled tube (lysates can be stored at –80 °C). For different time points, snap-freeze the lysates in liquid nitrogen instead of incubating it on ice.

5. Assay equal volume of extract (determined by cell number); keep some lysates to determine protein concentration for normalization of the relative fluorescence units (RFU) (although preferably, you determine it up front). Alternatively, it can use immobilized proteins purified by co-IP experiments using immobilized anti-NLRP7, such as from **step 10** of Subheading 3.3 in combination with SEC or directly from cell lysates, as source for the caspase-1. In this case, equilibrate beads in caspase assay buffer (lacking the substrate Ac-YVAD-AFC).

6. Turn on the fluorescent plate reader, set up the correct excitation and emission filters (AFC excitation, 400 nm; emission, 505 nm), and warm up the reader to 37 °C.

7. Prepare 100 μl caspase assay buffer/sample with fresh DTT and 100 μM of substrate Ac-YVAD-AFC.

8. Add same volume of adjusted lysates, including a negative control (max: 15 μl) into white or black 96-well plates.

9. Quickly add 100 μl of assay buffer to each well (if necessary remove air bubbles quickly with a syringe needle).

10. Measure caspase-1 activity over time for 1 h at 37 °C at 400/505 nm (excitation/emission) (*see* **Note 14**).

3.6 In Vitro Caspase-1 Activation Assay

1. Transfect a 100 mm tissue culture dish of HEK293 cells with a pro-IL-1β cDNA (*see* **Note 15**). 24–36 h post-transfection, lyse cells in caspase-1 lysis buffer, and centrifuge the lysates at high speed ($14,000 \times g$) for 10 min at 4 °C to obtain the cleared lysate (*see* **Note 16**).

2. Purify the NLRP7 complex and IgG control complex as described above (Subheading 3.3), except maintain the bound proteins on the Sepharose beads (do not elute proteins in Laemmli buffer) and equilibrate beads in caspase-1 assay buffer.

3. Incubate the Sepharose beads containing immobilized NLRP7 and IgG control complex in caspase-1 assay buffer with total cleared lysates from HEK293 for 1–2 h at 37 °C.

4. Stop reaction by adding 2× Laemmli buffer and analyzing the conversion of pro-IL-1β to mature IL-1β by Western blot (*see* **Note 17**).

3.7 Cross-Linking

1. Plate 4×10^6 cells in a 60 mm tissue culture plate.

2. Transfect or treat cells as described in previous Subheading 3.1 or 3.4.

3. Remove the culture supernatants and rinse the cells with ice-cold PBS. Add 1 ml of ice-cold PBS to the plate and remove the cells using a cell scraper and transfer to 1.5 ml microcentrifuge tubes.

4. Centrifuge the cells at $400 \times g$ for 10 min and discard the supernatant.

5. Add 500 μl ice-cold lysis buffer to the cell pellet and lyse the cells by shearing 10 times through a 21-gauge needle.

6. Reserve 50 μl of cell lysate for Western blot analysis.

7. Centrifuge the remaining lysates at $2500 \times g$ for 10 min at 4 °C.

8. Transfer the supernatants to fresh tubes. Resuspend the pellets in 500 μl PBS. Add 2 mM disuccinimidyl suberate (DSS) (from a **freshly prepared** 100 mM stock) to the resuspended pellets and the supernatants. Incubate at RT for 30 min with agitation on a rotator or nutator.

9. Centrifuge samples at $2500 \times g$ for 10 min at 4 °C.

10. Remove supernatants and quench the cross-linking by resuspending the cross-linked pellets in 60 μl Laemmli sample buffer.

11. Boil the samples for 10 min at 95 °C and analyze by running the samples on a SDS/PAGE gel and detect the respective NLR or ASC by Western blotting (*see* **Note 18**).

4 Notes

1. This approach can be modified with application of a reversible cross-linker, such as dithiobis(succinimidyl propionate) (DSP) or Lomant's reagent, which contains a disulfide bond in its spacer arm, which is cleaved by reducing agents, such as Laemmli buffer. This approach allows co-IP experiments under stringent conditions of cross-linked protein complexes and

detection of purified proteins by immunoblot according to their monomeric molecular weight.

2. This time point is too long for purifying the complex by co-IP and SEC or to determine caspase-1 activity, but works well for cross-linking. We used vaccinia virus or MCMV infection for shorter times (90 min) to determine the AIM2–ASC complex in response to viral DNA [7].

3. We use activation of NLRP3 by LPS/ATP and LPS/nigericin and AIM2 by poly(dA:dT) as examples. However, any other inflammasome activator can be used. Crude LPS can be substituted for ultra pure LPS, but it already causes some inflammasome activation. The timing can be adjusted as needed from as short as 1 h to overnight. Similarly, the concentration of ATP and nigericin can be adjusted to cause the appropriate level of activation (usually 3–5 mM ATP and 2–5 µM nigericin is sufficient). Any transfection reagent of choice can be used for poly(dA:dT) transfection. However, the concentration may need to be adjusted based on transfection efficiency and manufacturer's protocol. Poly(dA:dT) conjugated to the cationic lipid transfection reagent LyoVec is sold by InvivoGen for direct cytosolic delivery, but we did not obtain sufficient inflammasome activation in our hands.

4. The large number of cells is required to detect low expressing NLRs, when separated into 24 fractions.

5. Do not overdry the pellets or incubate at higher temperatures, as this will cause problems when resuspending the protein pellet. Incubating at room temperature is also sufficient, but may require longer incubation times.

6. In case the Laemmli buffer changes color to yellow (indicating leftover TCA and insufficient washing), neutralize pH with a drop of 1 M Tris base.

7. You will need more gels, or a wide gel (for example the Bio-Rad Criterion gels can run up to 26 samples).

8. Less cells can be used, but to achieve sensitivity we usually use 60 or 100 mm dishes.

9. For the NLRP3–ASC interaction, we commonly IP with an anti-ASC antibody (Santa Cruz, sc-22514-R) and detect NLRP3 with an anti-NLRP3 antibody (Adipogen, Cryo-2, AG-20B-0014). This NLRP3 antibody also works well for IP and cross-reacts with human and mouse NLRP3. We purify the NLRP7 complex with an anti-NLRP7 antibody (Imgenex, IMG-6357A).

10. This is the step to continue to the caspase-1 activity assay, rather than Western blot analysis. Alternatively, the beads can be divided for both analyses.

11. We established cells stably expressing myc-ASC (HEK293[ASC]) for this purpose and only transfect with the NLR of interest. This helps with the usually poor expression of NLRs. At minimum, adjust expression of ASC and NLRs by transfecting approximately 1/3 ASC and 2/3 NLR. The GFP fusion stabilizes expression of NLRP3 and NLRP7, but additionally we tested HA and Flag tags, which also work.

12. In the case of epitope-tagged proteins, consider pulling down with directly Sepharose-immobilized antibodies, which are widely available and used for detecting directly HRP-conjugated anti-tag antibodies, as ASC runs very close to the antibody light chain band.

13. Keep cells on ice all the time to minimize activation of caspase-1, which can be activated by cell lysing at 30 °C [10].

14. Prolonged incubation times for several hours may be necessary for diluted lysates or low activity.

15. We use HEK293 cells for this assay, as these cells do not express endogenous caspase-1 and therefore pro-IL-1β is not cleaved and is maintained as pro-IL-1β in the lysate. Any other cell type lacking caspase-1 would be equally suited.

16. Alternatively, a pro-IL-18 cDNA could be used, as pro-IL-18 is also cleaved by caspase-1 [13–15, 22]. Several other substrates are less well characterized [23–25]. Helpful is the use of a C-terminally epitope-tagged cDNA, as the N-terminal pro-domain is proteolytically removed by caspase-1.

17. The size of pro-IL-1β is 31 kDa and of mature IL-1β is 17.5 kDa. If you use IL-18, pro-IL-18 is 24 kDa and mature IL-18 is 18 kDa.

18. Consider running a gradient SDS/PAGE to properly resolve the multiple-sized oligomers.

Acknowledgments

This work was supported by the National Institutes of Health (GM071723, AI099009, AI120618 and AR064349 to C.S., AI120625 to C.S. and A.D. and AR066739 to A.D.) and the American Heart Association (12GRNT12080035 to C.S.).

References

1. Barbé F, Douglas T, Saleh M (2014) Advances in Nod-like receptors (NLR) biology. Cytokine Growth Factor Rev 25:681–697

2. Khare S, Luc N, Dorfleutner A, Stehlik C (2010) Inflammasomes and their activation. Crit Rev Immunol 30:463–487

3. Phizicky EM, Fields S (1995) Protein–protein interactions: methods for detection and analysis. Microbiol Rev 59:94–123

4. Miernyk JA, Thelen JJ (2008) Biochemical approaches for discovering protein–protein interactions. Plant J 53:597–609

5. Dwane S, Kiely PA (2011) Tools used to study how protein complexes are assembled in signaling cascades. Bioeng Bugs 2:247–259

6. Fekete S, Beck A, Veuthey J-L, Guillarme D (2014) Theory and practice of size exclusion chromatography for the analysis of protein aggregates. J Pharm Biomed Anal 101C:161–173

7. Khare S, Dorfleutner A, Bryan NB, Yun C et al (2012) An NLRP7-containing inflammasome mediates recognition of microbial lipopeptides in human macrophages. Immunity 36:464–476

8. Kaboord B, Perr M (2008) Isolation of proteins and protein complexes by immunoprecipitation. Methods Mol Biol 424:349–364

9. Markham K, Bai Y, Schmitt-Ulms G (2007) Co-immunoprecipitations revisited: an update on experimental concepts and their implementation for sensitive interactome investigations of endogenous proteins. Anal Bioanal Chem 389:461–473

10. Martinon F, Burns K, Tschopp J (2002) The inflammasome: a molecular platform triggering activation of inflammatory caspases and processing of proIL-beta. Mol Cell 10:417–426

11. Elliott JM, Rouge L, Wiesmann C, Scheer JM (2009) Crystal structure of procaspase-1 zymogen domain reveals insight into inflammatory caspase autoactivation. J Biol Chem 284:6546–6553

12. Yamin TT, Ayala JM, Miller DK (1996) Activation of the native 45-kDa precursor form of interleukin-1-converting enzyme. J Biol Chem 271:13273–13282

13. Kuida K, Lippke JA, Ku G, Harding MW et al (1995) Altered cytokine export and apoptosis in mice deficient in interleukin-1 beta converting enzyme. Science 267:2000–2003

14. Li P, Allen H, Banerjee S, Franklin S et al (1995) Mice deficient in IL-1b-converting enzyme are defective in production of mature IL-1b and resistant to endotoxic shock. Cell 80:401–411

15. Thornberry NA, Bull HG, Calaycay JR, Chapman KT et al (1992) A novel heterodimeric cysteine protease is required for interleukin-1 beta processing in monocytes. Nature 356:768–774

16. Thornberry NA, Rano TA, Peterson EP, Rasper DM et al (1997) A combinatorial approach defines specificities of members of the caspase family and granzyme B. Functional relationships established for key mediators of apoptosis. J Biol Chem 272:17907–17911

17. Mattson G, Conklin E, Desai S, Nielander G et al (1993) A practical approach to crosslinking. Mol Biol Rep 17:167–183

18. Fernandes-Alnemri T, Wu J, Yu JW, Datta P et al (2007) The pyroptosome: a supramolecular assembly of ASC dimers mediating inflammatory cell death via caspase-1 activation. Cell Death Differ 14:1590–1604

19. Fernandes-Alnemri T, Alnemri ES (2008) Assembly, purification, and assay of the activity of the ASC pyroptosome. Methods Enzymol 442:251–270

20. Khare S, Ratsimandresy RA, de Almeida L, Cuda CM et al (2014) The PYRIN domain-only protein POP3 inhibits ALR inflammasomes and regulates responses to infection with DNA viruses. Nat Immunol 15:343–353

21. Masters SC (2004) Co-immunoprecipitation from transfected cells. Methods Mol Biol 261:337–350

22. Ghayur T, Banerjee S, Hugunin M, Butler D et al (1997) Caspase-1 processes IFN-gamma-inducing factor and regulates LPS-induced IFN-gamma production. Nature 386:619–623

23. Gurcel L, Abrami L, Girardin S, Tschopp J, van der Goot FG (2006) Caspase-1 activation of lipid metabolic pathways in response to bacterial pore-forming toxins promotes cell survival. Cell 126:1135–1145

24. Keller M, Ruegg A, Werner S, Beer HD (2008) Active caspase-1 is a regulator of unconventional protein secretion. Cell 132:818–831

25. Shao W, Yeretssian G, Doiron K, Hussain SN, Saleh M (2007) The caspase-1 digestome identifies the glycolysis pathway as a target during infection and septic shock. J Biol Chem 282:36321–36329

Chapter 9

Measuring NLR Oligomerization II: Detection of ASC Speck Formation by Confocal Microscopy and Immunofluorescence

Michael Beilharz, Dominic De Nardo, Eicke Latz, and Bernardo S. Franklin

Abstract

Inflammasome assembly results in the formation of a large intracellular protein scaffold driven by the oligomerization of the adaptor protein apoptosis-associated speck-like protein containing a CARD (ASC). Following inflammasome activation, ASC polymerizes to form a large singular structure termed the ASC "speck," which is crucial for recruitment of caspase-1 and its inflammatory activity. Hence, due to the considerably large size of these structures, ASC specks can be easily visualized by microscopy as a simple upstream readout for inflammasome activation. Here, we provide two detailed protocols for imaging ASC specks: by (1) live-cell imaging of monocyte/macrophage cell lines expressing a fluorescently tagged version of ASC and (2) immunofluorescence of endogenous ASC in cell lines and human immune cells. In addition, we outline a protocol for increasing the specificity of ASC antibodies for use in immunofluorescence.

Key words Inflammasome, ASC, Speck, Live-cell imaging, Immunofluorescence, Confocal microscopy, Flow cytometry

1 Introduction

The activation of certain innate immune receptors by microbial and endogenous danger signals leads to the formation of large multi-protein signaling platforms called inflammasomes. The assembly of inflammasomes mediates caspase-1-dependent proteolytic cleavage of important inflammatory mediators of the IL-1 family of cytokines, IL-1β, and IL-18, in concert with pyroptotic cell death [1]. In addition to their intracellular function, inflammasomes have recently been shown to display extracellular activity following their release from cells undergoing pyroptosis [2, 3]. Inflammasomes typically consist of three major components, the

The original version of this chapter was revised. The erratum to this chapter is available at: DOI 10.1007/978-1-4939-3289-4_19

Francesco Di Virgilio and Pablo Pelegrín (eds.), *NLR Proteins: Methods and Protocols*, Methods in Molecular Biology, vol. 1417, DOI 10.1007/978-1-4939-3566-6_9, © Springer Science+Business Media New York 2016

receptor/sensor (e.g., NLRP3, AIM2), the adaptor ASC, and the effector caspase-1. The association between sensor and ASC molecules is mediated via pyrin domain (PYD) interactions, while the involvement of caspase-1 is dependent on a caspase recruitment domain (CARD) present on both ASC and caspase-1. The activation of the inflammasome sensors triggers the recruitment and prion-like polymerization of ASC into large filamentous structures that triggers the inflammatory activity of caspase-1 [4, 5]. In resting cells, ASC displays soluble cytoplasmic and nuclear localization, but, upon assembly of inflammasomes, it is mobilized to form a large singular paranuclear ASC "speck" (~1 µm in diameter) [2, 6, 7]. Due to their relatively large size, ASC specks are easily visualized by various imaging techniques and therefore represent a convenient readout for inflammasome activation. We and others have previously described protocols for the generation of ASC fluorescently tagged cell lines using a retroviral transduction system and their use in high-throughput imaging for quantification and assessment of ASC speck formation following inflammasome activation [8, 9]; furthermore, methods to generate ASC specks from recombinant sources or purify from activated cells are also described [2, 7, 9].

In this chapter, we describe two imaging-based techniques for the detection of ASC specks. Firstly, we provide a protocol to assess the localization of ASC fluorescently tagged in live cells over time following NLRP3 inflammasome activation using confocal microscopy. Secondly, we present an optimized protocol for performing immunofluorescence (IF) of ASC in cell lines and primary human immune cells with commercially available antibodies. Finally, we provide a method for increasing the specificity of ASC antibodies to use in immunofluorescence by preclearance of anti-ASC antibodies using macrophages deficient in ASC ($Asc^{-/-}$ iMΦs).

2   Materials

2.1   Cells and Tissue Culture

1. THP-1 monocytes, or mouse-immortalized macrophages (iMΦs) expressing ASC fused to a fluorescent protein (e.g., cyan fluorescent protein, CFP)

2. THP-1 cells (American-type culture collection- ATCC), human monocyte cell line

3. Human peripheral blood mononucleated cells (PBMCs)

4. Wild-type bone marrow-derived macrophages (BMDMs)

5. ASC-deficient ($Asc^{-/-}$) BMDMs

6. Ficoll

7. Tissue culture flasks, 75 and 175 cm²

8. 96-well tissue culture-treated plates for imaging (with flat glass bottom)

9. "Complete" Dulbecco's modified Eagle's medium (DMEM) cell culture medium: DMEM, supplemented with 10 % fetal bovine serum (FBS); optional: antibiotics (e.g., 10 μg/ml ciprofloxacin, or 100 U/mL of Penicillin-Streptomycin)

10. "Complete" Roswell Park Memorial Institute (RPMI) cell culture medium: RPMI supplemented with 10 % FBS; optional: antibiotics (e.g., 10 μg/ml ciprofloxacin, or Penicillin-Streptomycin)

11. Phosphate-buffered saline solution (PBS): 137 mM NaCl, 2.7 mM KCl, 1 mM Na_2HPO_4, and 2 mM KH_2PO_4 and pH 7.4

12. Mammalian cell culture facility with an incubator maintained at 37 °C and 5 % CO_2 and 95 % air

2.2 Stimuli

1. Phorbol 12-myristate 13-acetate (PMA)

2. Ultrapure lipopolysaccharide from *Escherichia coli* 0111:B4 (LPS)

3. Nigericin

4. Adenosine 5′-triphosphate disodium salt hydrate (ATP) (*see* **Note 1**)

2.3 Immunostaining and Flow Cytometry

1. Sixteen percent formaldehyde (store protected from light)

2. Fixation buffer: 4 % formaldehyde in PBS (*see* **Note 2**)

3. Permeabilization and blocking (perm/block) buffer: 10 % goat serum, 1 % FBS, and 0.5 % Triton-X100 in PBS

4. Wash buffer: PBS

5. DRAQ5 (store at 4 °C protected from light; DRAQ5 is toxic, so handle with care)

6. Recombinant cholera toxin subunit B-conjugated with Alexa Fluor 555® (this product is exclusive of Life Technologies, although other fluorochrome conjugates from other companies are also suitable). Fluorochrome-conjugated Weat Germ Agglutinin also work well and can be used as an alternative cell membrane staining.

2.4 Antibodies and Conjugates

1. There are different anti-ASC antibodies suitable to detect ASC; here we use the rabbit polyclonals (pAb) anti-ASC from Adipogen (clone AL177) and Santa Cruz (Clone N-15-R) and the mouse monoclonals (mAbs) anti-ASC from Millipore (clone 2EI-7), and BioLegend (purified TMS-1 antibody, clone HASC-71), although other clones might be also suitable.

2. Mouse-purified IgG isotype control antibody

3. Rabbit-purified IgG isotype control antibody

4. Goat anti-rabbit-Alexa Fluor 488® and anti-mouse-Alexa Fluor 488® secondary antibodies (these antibodies are exclusive of Life Technologies, although other fluorochrome conjugated antibodies from other companies are also suitable)

2.5 Imaging and Flow Cytometry

1. Epifluorescence microscope with a filter for CFP (e.g., 458 nm laser with 510/80 nm emission optics) and a filter for DRAQ5 (e.g., Filter Set 50 from Zeiss or similar: excitation band pass 640/30, dichroic mirror FT 660, emission band pass 690/50) and with 20× and 63× objective. As an alternative, a confocal microscope with laser lines 458 nm (or 405 nm) and 647 nm (or 633 nm) and appropriate filters (as above) can be used.

2. Image analysis software such as Volocity 6.01 or ImageJ (available as freeware from http://rsbweb.nih.gov/ij).

3. Flow cytometer equipped with argon laser and filter settings for CFP (e.g., 445 nm laser with 510/80 nm emission optics) and GFP/FITC (e.g., 488 nm laser with 505/45 nm emission optics).

4. Flow cytometry software such as FlowJo V9.8 or later.

3   Methods

3.1 Live Imaging of ASC Speck Formation

This method is suitable to study ASC speck formation by live imaging upon NLRP3 inflammasome activation in cells stably expressing fluorescent ASC.

3.1.1 Stimulation and Counterstaining for Live Imaging

1. **Optional**: To minimize cell lost during the staining protocol, it is recommended to coat the wells of 96-well imaging plates with either poly-L-lysine or mouse collagen IV before seeding the cells (*see* **Note 3**).

2. Seed 0.5×10^5 iMΦs in complete DMEM or 1×10^5 THP-1 monocytes in complete RPMI (supplemented with 100 nM of PMA) per well of a 96-well imaging plate (*see* **Note 4**).

3. Incubate the cells overnight at 37 °C and 5 % CO_2.

4. The next day, prime cells with 200–250 ng/ml of LPS for iMΦs or 1 μg/ml of LPS for THP-1 for 2–3 h in 100 μl of complete media (use DMEM for iMΦs and RPMI for THP-1).

5. If nuclei and/or membrane staining is desired, it can be done at this stage. During the last 30 min of LPS priming, DRAQ5 or cholera toxin subunit B (CtxB) can be added to the cells. *See* **Note 5**.

6. Place the plate containing the cells in the imaging stage in the microscope and adjust the laser intensities, the focal plane, and the desired field for imaging. *See* **Note 6**.

7. Remove the media containing LPS, and replace it with 100 µl of fresh complete media containing NLRP3 inflammasome activators (10 µM of nigericin or 5 mM ATP).

8. Place the plate in a confocal microscope equipped with an incubation chamber set at 37 °C and 5 % CO_2 and start imaging (see **Note 7**).

3.1.2 Confocal Settings and Image Acquisition

9. Use the 20× or 63× objective with filters for the fluorescent tag of your ASC construct (i.e., we use a laser with 458 nm [or 405 nm] wavelength to excite ASC-CFP, 561 nm to excite CtxB-Alexa 555, and one with 647 nm [or 633 nm] to excite DRAQ5).

10. During acquisition, make sure to use the same imaging parameters (gain, light intensity, pinhole, etc.) for all conditions.

11. The formation of ASC specks is rapid, so adjust the interval between every picture accordingly. We recommend following the assembly of ASC fluorescent protein by acquiring images every 5 min for up to 90 min total imaging time.

12. While imaging the fluorescently tagged version of ASC, you should see a dramatic change in its cellular distribution. The fluorescent ASC changes from a weak diffuse signal present throughout the cell (including the nucleus) to a singular bright spot in an activated cell (Fig. 1a).

3.2 Immunofluorescence of ASC Specks with Antibodies

ASC specks can also be targeted by antibodies immediately upon inflammasome activation [2] (see **Note 7** and Fig. 1b).

We have tested several commercially available anti-ASC antibodies, and their combination with secondary antibodies directly conjugated to Alexa 488, and found significant differences in their performance (e.g., signal/noise and specificity) (see Fig 1b, **Note 8–9,** and Fig. 2). Irrespective of the anti-ASC antibody of your choice, we provide here a general protocol to stain endogenous ASC in inflammasome-activated cells:

1. *Optional*: To minimize cell lost during the staining protocol, it is recommended to coat the wells of 96-well imaging plates with either poly-L-lysine or mouse collagen IV before seeding the cells (see **Note 3**).

2. Seed wild-type THP-1 monocytes, iMΦs, or BMDMs in 96-well confocal plates, prime cells with LPS, and activate the NLRP3 inflammasome with nigericin or ATP as described in the previous section.

3. Upon activation (60 min for ATP and 90 min for nigericin), carefully remove the media on the cells and slowly add 100 µl of the fixation buffer and incubate samples for 30 min at 37 °C (see **Note 10**).

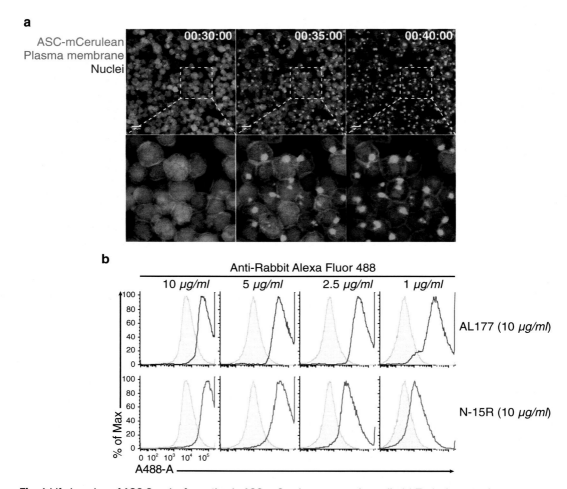

Fig. 1 Life imaging of ASC Specks formation in ASC-mCerulean expressing cells (**a**) Typical result of unstimulated and LPS-primed, nigericin-activated ASC-mCerulean-expressing human THP-1 monocytes. Images were acquired in a Leica TCS SMD FLCS confocal microscope (63× objective). Maximum projection images are shown. Note the redistribution of ASC-mCerulean from a broad and dim fluorescence in the *left panels* to a bright, small speck-like fluorescence in the *right panels*. Nuclei were stained with DRAQ5 (5 µM) and plasma membrane with cholera toxin B-Alexa Fluor 555 (5 µg/mL). *White numbers top-right corners* are imaging time. (**b**) Titration of the secondary anti-species IgG specific antibodies directly conjugated to Alexa Fluor 488 dye for detection of ASC specks by Flow Cytometry

4. Carefully wash the cells twice by slowly adding 100 µl PBS (*see* **Note 11**) to remove fixation buffer.

5. Slowly add 100 µl of the perm/block buffer to the cells and incubate for 30 min at 37 °C. This blocking step is important to reduce unspecific staining.

6. Slowly remove the perm/block buffer and replace it with 100 µl/well of fresh perm/block buffer containing 1 µg of the anti-ASC of your choice (*see* **Note 9** and Fig. 2).

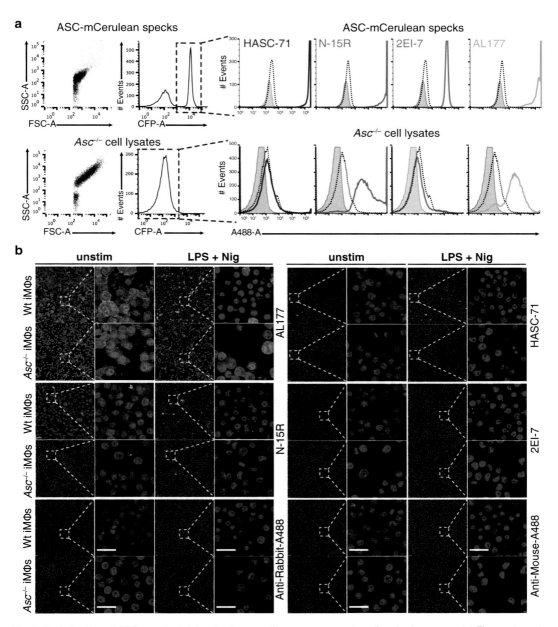

Fig. 2 Optimization of ASC speck staining for immunofluorescence and confocal microscopy (**a**) Flow cytometric analysis of ASC-mCerulean specks, or lysates of *Asc*[−/−] iMΦs, that were stained with a series of commercially available anti-ASC antibodies, followed by staining with 2.5 μg/ml of secondary antibodies directly conjugated to Alexa 488. Note that although all antibodies tested yielded a distinct staining of ASC-mCerulean specks, the antibodies from Santa Cruz clone N15-R (*red trace*) and Adipogen (*orange trace*) also reacted quite highly with *Asc*[−/−] iMΦs. (**b**) Confocal imaging of resting or LPS-primed, nigericin-activated wild-type or *Asc*[−/−] iMΦs. Cells were fixed and stained with a series of commercially available anti-ASC antibodies, followed by staining with secondary antibodies directly conjugated to Alexa 488. As a control cells were incubated with the secondary antibodies alone

7. Incubate cells at room temperature for at least 1 h (*see* **Note 12**).

8. Wash cells gently three times with 200 μl of block/perm buffer (*see* **Note 11**).

9. Stain cells with secondary antibody (2.5 μg/ml, Fig. 1b) in 100 μl of block/perm buffer and incubate for 1 h at RT.

10. Wash cells gently three times with 200 μl block/perm buffer.

11. If nuclear or plasma membrane staining is desired, add 5 μM of DRAQ5 and/or 5 μg/ml of CtxB in 200 μl PBS and incubate 5–10 min at 37 °C.

12. Wash the cells gently three times with 200 μl PBS (*see* **Note 11**).

13. Plates can be stored at 4 °C, with 100 μl of PBS per well until imaged.

14. Image cells with a fluorescent or confocal microscope, as described above in Subheading 3.1.2.

15. Similar to live imaging of fluorescent ASC-expressing cells, you should see a dramatic change in the distribution of antibody-targeted ASC within the cell (Fig. 3).

3.3 Imaging of ASC Specks Using Antibodies in PBMCs

1. Isolate human PBMCs using Ficoll according to manufacturer's instructions.

2. PBMCs adhere very weakly to confocal plates. To minimize cell lost during the staining protocol, it is recommended to coat the wells of 96-well imaging plates with either poly-L-lysine or mouse collagen IV before seeding the cells (*see* **Note 3**).

3. Seed 1×10^5 cells/well in a 96-well glass-bottom plate (*see* **Note 4**).

4. Let the cells adhere for at least 3 h (or overnight) at 37 °C.

5. Gently remove the media and wash the cells once with PBS.

6. Prime the cells with 200–1000 pg/ml LPS for 2–3 h in 100 μl of complete DMEM per well (*see* **Note 13**).

7. Remove the supernatants and wash the cells once with 100 μl of PBS to remove serum-containing medium.

8. Add 100 μl/well of serum-free medium containing 10 μM of nigericin or 5 mM of ATP and incubate cells at 37 °C, 5 % CO_2 for 30–60 min with ATP or 90 min with nigericin.

9. After inflammasome activation, centrifuge the plate at $400 \times g$ for 5 min to pellet floating cells and carefully remove the supernatants.

10. Fix cells with 100 μl of fixation buffer/well and incubate at 37 °C for 30 min (or overnight at 4 °C).

11. Wash the cells very carefully three times with PBS and incubate at 37 °C for 5 min between washes (*see* **Note 11**).

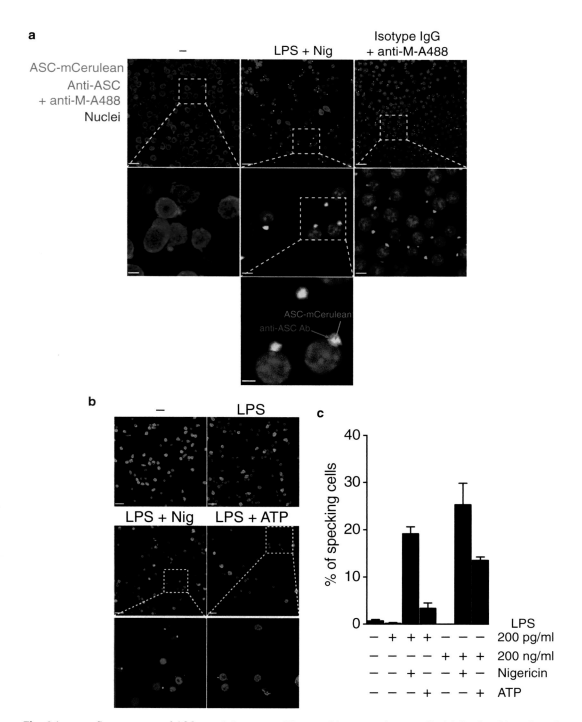

Fig. 3 Immunofluorescence of ASC speck in mouse iMΦs and human primary cells (**a**) Confocal imaging of resting or LPS-primed nigericin-activated ASC-mCerulean-expressing iMΦs. Cells were fixed and stained with anti-ASC (HASC-71, BioLegend) Abs, followed by staining with secondary anti-mouse IgG directly conjugated to Alexa Fluor 488 dye. As a control, cells were incubated with a mouse-purified IgG isotype, followed by staining with secondary antibody. Nuclei were stained with DRAQ5. (**b**) Confocal imaging of resting (−) or LPS-primed (LPS) human PBMCs that were either left untreated or activated with nigericin (LPS + Nig) or ATP (LPS + ATP). Cells were fixed and stained with anti-ASC (HASC-71, BioLegend) antibody, followed by staining with 2.5 μg/ml of anti-rabbit IgG directly conjugated to Alexa 488. (**c**) Quantification of ASC specks in human PBMCs treated and stained as in (**b**)

12. Incubate the cells with 100 μl of perm/block buffer for 30 min at 37 °C.

13. Add 100 μl of perm/block buffer containing 10 μg/ml of the anti-ASC of your choice in the desired wells and incubate plates at room temp for 1 h or overnight at 4 °C.

14. Carefully wash the wells three times with PBS.

15. Add 100 μl of perm/block buffer containing 2.5 μg/ml of anti-rabbit IgG Alexa Fluor 488 per well and incubate the plate at 37 °C for 40–60 min.

16. Carefully wash the wells three times with PBS.

17. If nuclear or plasma membrane staining is desired, add DRAQ5 and/or CtxB as described above.

18. Carefully wash the wells three times with PBS.

19. Add 100 μl of PBS to the cells and proceed to imaging.

20. Image cells by confocal microscopy as described above in Subheading 3.1.2.

3.4 Protocol to Increase the Specificity of Anti-ASC Antibodies for Immunofluorescence

Antibodies subjected to this method can be analyzed by flow cytometry to assess their binding capacity to purified ASC specks or $Asc^{-/-}$ cell lysates. As shown in Fig. 4, the Anti-ASC N-15R antibody (Santa Cruz) stains ASC-mCerulean specks; however, it also reacts with $Asc^{-/-}$ cell lysates (mock specks) yielding high staining background. However, such a nonspecific signal is not observed with the anti-ASC HASC-71 (BioLegend) antibody. It is thus possible to "clean" the anti-ASC N-15R antibody to significantly reduce nonspecific staining in $Asc^{-/-}$ cell lysates, without affecting the staining of ASC specks. This method may also be applied to other anti-ASC antibodies. Here we provide a protocol to "clean" anti-ASC antibodies:

1. Harvest ~1×10^6 $Asc^{-/-}$ cells/ml and transfer the cell suspension to 50 ml tubes.

2. Wash cells twice with cold PBS by pelleting them at $400 \times g$ for 5 min at 4 °C.

3. Resuspend cells in 10 ml of block/perm buffer and incubate them for 30 min at room temperature.

4. Wash cells twice by pelleting them at $400 \times g$ for 5 min at 4 °C and resuspending in PBS.

5. Resuspend cells in 1 ml of block/perm buffer containing 20 μg of Anti-ASC N-15R antibody.

6. Incubate cells overnight at 4 °C with rotation, to allow the antibody to evenly diffuse.

7. The next day, pellet the cells at $400 \times g$ for 5 min and transfer the supernatants containing unbound antibodies to a new tube.

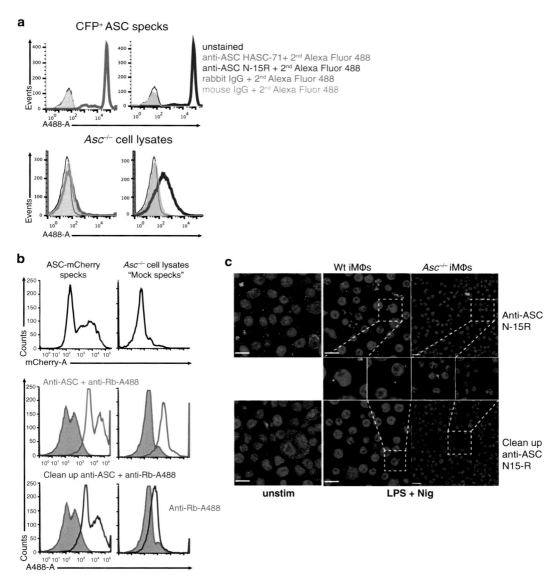

Fig. 4 Optimization of anti-ASC performance for immunofluorescence. (**a**) Flow cytometry of ASC-mCerulean specks or lysates of *Asc*[−/−] iMΦs stained with anti-ASC HASC-71 (BioLegend, *red trace*) or N-15R (Santa Cruz, *blue trace*), followed by staining with 2.5 μg/ml of anti-mouse or anti-rabbit IgG directly conjugated to Alexa Fluor 488 dye. As a control, cells were incubated with a mouse- or rabbit-purified IgG isotype, followed by staining with secondary antibody. Note that although both antibodies yield a staining of ASC-mCerulean specks, the Santa Cruz antibody also reacts quite highly with *Asc*[−/−] iMΦs. (**b**) Flow cytometry analysis as in a of ASC-mCerulean specks or lysates of *Asc*[−/−] iMΦs stained with anti-ASC from Santa Cruz before (*red trace*) of after (*blue trace*) cleanup of this antibody, followed by staining with 2.5 μg/ml of anti-rabbit IgG directly conjugated to Alexa 488. (**c**) The comparison of the anti-ASC Santa Cruz abs in resting (unstim) or LPS-primed, nigericin-activated (LPS + Nig) wild-type iMΦs before and after cleanup. Cells were fixed and stained with anti-ASC antibodies, followed by staining with 2.5 μg/ml of anti-rabbit IgG directly conjugated to Alexa 488

8. Filter the supernatants at 0.22 μm to remove any cell debris.

9. Determine the protein concentration in the supernatants (e.g., by BCA or Bradford).

10. Test the clean anti-ASC antibody in flow cytometry or confocal assays (Fig. 4).

4   Notes

1. Many inflammasome stimuli are toxic and need to be handled with care. Follow the manufacturer's instructions. To activate NLRP3, the following conditions can be used: prime cells with 200 ng/ml LPS for 2–3 h, and then stimulate with 5 mM ATP for 30–60 min or 10 μM nigericin for 60–90 min.

2. Formaldehyde is toxic and carcinogenic. Wear protective equipment and dispose of formaldehyde according to the local regulations. Formaldehyde solutions need to be made up fresh every time, as formaldehyde oxidizes to formic acid over time.

3. As inflammasome activation leads to pyroptotic cell death, cells often detach from the plate and can therefore be washed away during the washing steps, leaving very few cells for imaging. This can be minimized by pre-coating the plates with either 30–40 μl/well of poly-L-lysine for 1 h, or with 1–5 μg/ml of mouse collagen IV for 30 min at 37 °C. Wash the plates with sterile injection water and let them dry completely before seeding the cells.

4. It is advised to not use the outer wells of the 96-well plate, as they are more prone to evaporation. Thus, only use the middle 60 wells from columns 2 to 10 and from rows B to G.

5. For live imaging, DRAQ5 and/or cholera toxin B (CtxB) can be added to the cells during the final 30 min of LPS priming. Dilute DRAQ5 or CtxB in PBS and add to the cells at a concentration of 2.5–5.0 μM for DRAQ5 and 5 μg/ml for CtxB. Incubate for 5–10 min at 37 °C. Wash it once with 100 μl/well of PBS and replace the PBS with fresh medium. Follow with inflammasome stimulation.

6. Once the inflammasome stimuli are added, cells are rapidly activated. Therefore, it is recommended for optimal imaging performance to pre-image the plates before activation. This can be done during the last 30 min of the LPS priming step. Place the plate containing the cells in the imaging stage in the microscope and adjust the laser intensities, the focal plane, and the desired field for imaging. If stacks in the z-axis are desired, this can also be adjusted at this point. This procedure will spare the time needed to set up the microscope and guarantee that the cells are imaged from the very initial stages of activation.

7. Inflammasome activation in cells usually results in disintegration of the plasma membrane and cell morphology. Cells will often blow and round up. It is thus recommended to take z-stacks during imaging and display the maximum intensity projection of these z-stacks. Cells with specks will often round up, and some of the specks will be located further away from the focal plane.

8. To optimize staining of ASC specks for immunofluorescence, it is important to titrate the antibodies to get the best signal/noise ratio. We found out that most primary anti-ASC antibodies work well at a concentration of 10 μg/ml, thought lower concentrations also work. We have tested different concentrations of the fluorescently conjugated secondary antibodies and found that the optimal signal-to-noise ratio was obtained when secondary antibodies were used at a final concentration of 2.5 μg/ml (Fig. 1b). Unless stated otherwise, this concentration was used throughout in the experiments described in this book chapter.

9. We have tested most of the commercially available anti-ASC antibodies by flow cytometry and confocal imaging (Fig. 2), in different concentrations and conditions (not shown). In our laboratory conditions, the mouse monoclonal anti-ASC from BioLegend and Millipore displayed the best signal/noise ratio and were highly specific, as they did not react with $Asc^{-/-}$ cell lysates (Fig. 2). Although all antibodies tested yielded a distinct staining of ASC specks, the antibodies from Santa Cruz and Adipogen displayed less specificity as they reacted quite highly with $Asc^{-/-}$ iMΦs (Fig. 2a, traces of red and orange for Santa Cruz and Adipogen antibodies, respectively).

10. Is it recommended to keep all buffers (fixation and block/perm) cold.

11. It is crucial at this point to wash the wells very gently to avoid loss of cells from the plate.

12. Incubation with anti-ASC antibodies overnight at 4 °C usually yields better signal.

13. Human PBMCs can be primed with 200 pg/ml or 1 ng/ml of LPS. In our hands, 1 ng/ml resulted in a higher percentage of inflammasome-activated cells (as assessed by quantifying the cells that contains ASC specks, Fig. 3c).

Acknowledgments

This work was supported by grants from the Alexander von Humboldt Foundation (B.S.F.) and the intramural BONFOR research support at the University of Bonn (B.S.F. and D.D).

References

1. Latz E et al (2013) Activation and regulation of the inflammasomes. Nat Rev Immunol 13:397–411
2. Franklin BS et al (2014) The adaptor ASC has extracellular and "prionoid" activities that propagate inflammation. Nat Immunol 15:727–737. doi:10.1038/ni.2913
3. Baroja-Mazo A et al (2014) The NLRP3 inflammasome is released as a particulate danger signal that amplifies the inflammatory response. Nat Immunol 15(8):738–748
4. Cai X et al (2014) Prion-like polymerization underlies signal transduction in antiviral immune defense and inflammasome activation. Cell 156:1207–1222
5. Lu A et al (2014) Unified polymerization mechanism for the assembly of ASC-dependent inflammasomes. Cell 156:1193–1206
6. Masumoto J et al (1999) ASC, a novel 22-kDa protein, aggregates during apoptosis of human promyelocytic leukemia HL-60 cells. J Biol Chem 274:33835–33838
7. Fernandes-Alnemri T et al (2007) The pyroptosome: a supramolecular assembly of ASC dimers mediating inflammatory cell death via caspase-1 activation. Cell Death Differ 14:1590–1604
8. Stutz A, Horvath GL, Monks BG, Latz E (2013) ASC speck formation as a readout for inflammasome activation. Methods Mol Biol 1040:91–101
9. Fernandes-Alnemri T, Alnemri ES (2008) Assembly, purification, and assay of the activity of the ASC pyroptosome. Methods Enzymol 442:251–270

Chapter 10

Measuring NLR Oligomerization III: Detection of NLRP3 Complex by Bioluminescence Resonance Energy Transfer

Fátima Martín-Sánchez, Vincent Compan, and Pablo Pelegrín

Abstract

Bioluminescent resonance energy transfer (BRET) is a natural phenomenon resulting from a non-radiative energy transfer between a bioluminescent donor (*Renilla* luciferase) and a fluorescent protein acceptor. BRET signal is dependent on the distance and the orientation between the donor and the acceptor and could be used to study protein–protein interactions and conformational changes within proteins in real time in living cells. This protocol describes the use of BRET technique to study NLRP3 oligomerization in living cells before and during NLRP3 inflammasome activation.

Key words BRET, NLRP3, Sensor, Bioluminescence

1 Introduction

BRET assay is an easy methodological tool used to study protein–protein interactions in real time in living cells [1–3]. BRET requires tagging the proteins of interest with a bioluminescent donor molecule (*Renilla* luciferase or rLuc) and a fluorescent acceptor molecule, and their co-expression into suitable host cells. BRET procedure implies an energy transfer from the donor to the acceptor, which reemits light at a different wavelength. Energy transfer occurs when donor and acceptor are in close proximity, less than 100 Å and requires an overlap between the emission spectrum of the donor and the absorption spectrum of the acceptor [1, 4, 5].

Depending on the rLuc substrate used (for example coelenterazine-h, DeepBlueC™ or a protected form of coelenterazine-h, termed EnduRen™), and the nature of the acceptor (green fluorescent protein variants or quantum dot), different generations of BRET have been developed [1]. The main donor and acceptor used in BRET are a variant of *Renilla* luciferase (rLuc) and the yellow fluorescent protein (YFP) respectively, and require the substrate coelenterazine-h in case of BRET1. The oxidation of the substrate by the rLuc results in light emission sufficient to

Francesco Di Virgilio and Pablo Pelegrín (eds.), *NLR Proteins: Methods and Protocols*, Methods in Molecular Biology, vol. 1417, DOI 10.1007/978-1-4939-3566-6_10, © Springer Science+Business Media New York 2016

No protein-protein interaction

Protein-protein interaction

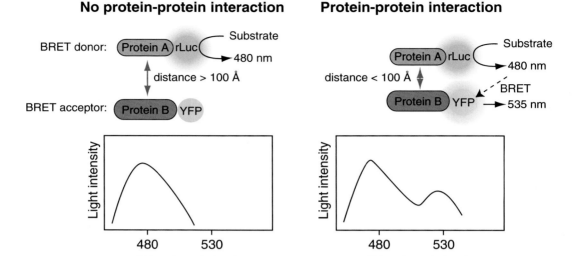

Fig. 1 Diagram illustrating Bioluminescent Resonance Energy Transfer (BRET). Schematic representation of BRET1 between two different proteins tagged with the donor molecule rLuc and the acceptor molecule YFP. When the proteins are at less than 100 Å, energy transfer can occur and the acceptor molecule emits

excite the YFP, which then emits fluorescence at a different wavelength (Fig. 1). The energy transferred from the donor to the acceptor is detected and can be measured as BRET signal [1, 6]. BRET has some advantages over FRET since it does not require an external light source to initiate the energy transfer, minimizing autofluorescence, photobleaching of the acceptor, and possible cell photodamage.

Here, we focus on the use of two models (model-A and model-B) of BRET1 assay to study oligomerization and conformational changes of the Nod-like receptor (NLR) with a pyrin domain 3 (NLRP3) in living cells before and during inflammasome activation (Fig. 2).

In model-A, we examined intermolecular interactions between NLRP3 proteins by using two NLRP3 constructs, one fused to rLuc on the C-terminus and one fused to the YFP on the N-terminus. Using the model-A, we have proposed that previous to stimulation, NLRP3 proteins expressed in HEK293 cells are in spatial proximity and form a preassembled complex, as indicated by a specific basal BRET signal (Fig. 2). This initial conformation is altered during inflammasome activation by hypotonicity or nigericin treatment of the cells [7].

In model-B, we produced a NLRP3 protein fused to rLuc on the C-terminus and the YFP to the N-terminus to record intramolecular BRET variations (Fig. 2). This model is useful to study NLRP3 conformational changes during inflammasome activation. In resting conditions the conformation of NLRP3 allows an intramolecular energy transfer between the donor and the acceptor, as

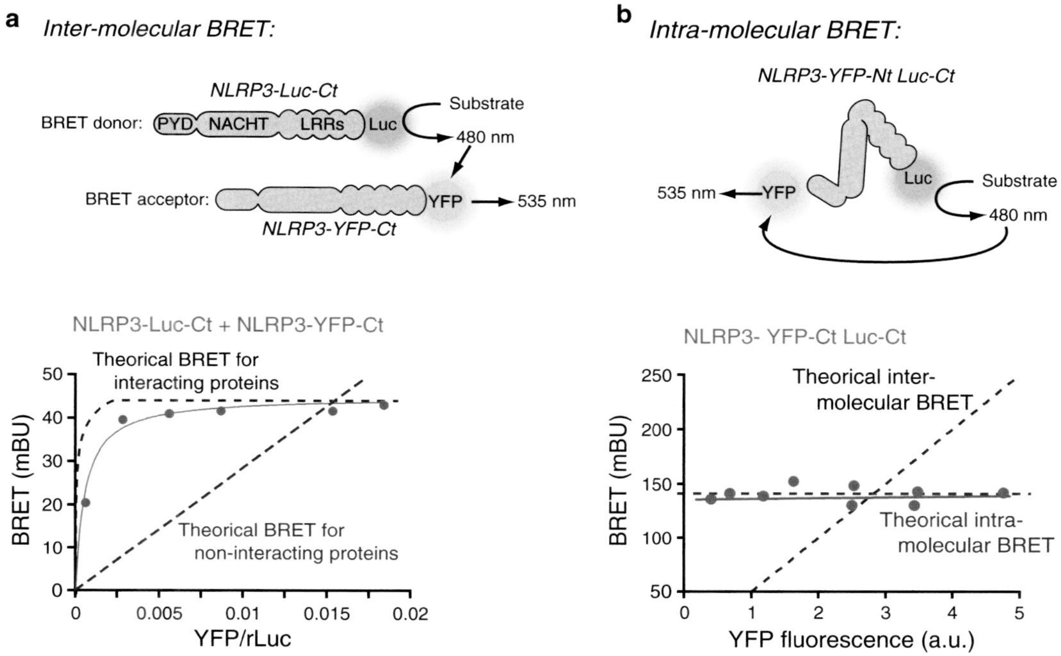

Fig. 2 Diagram illustrating two approaches to measure BRET on NLRP3 molecules. (**a**) In model-A specific intermolecular BRET was studied using two NLRP3 molecules, one fused to rLuc acting as a donor, and one fused to YFP acting as acceptor. Specific BRET among NLRP3 molecules can be detected when increasing concentrations of acceptor results in a saturation curve (*red trace*). (**b**) In model-B NLRP3 intramolecular BRET was studied using one NLRP3 molecule fused to rLuc and YFP. In case of intramolecular BRET, a constant BRET signal is obtained even when increasing concentrations of the NLRP3 sensor are transfected (*red trace*)

indicated by a stable BRET signal (Fig. 2). During the initial inflammasome activation, NLRP3 protein conformational changes will be monitored with a variation of the energy transfer.

In both models the constructs were transfected and expressed in HEK293 cells. The rLuc substrate used was coelenterazine-h, which oxidation emits light with a single peak at 480 nm, exciting the YFP acceptor that then emits at 535 nm. As a signal control, we used HEK293 cells expressing the construct NLRP3-rLuc-Ct (donor alone). The BRET assays were performed on a multiplate reader to measure the light emitted at 480 and 535 nm and BRET signal is expressed as milliBRET units (mBU).

2 Materials

Prepare all solutions using ultrapure water and analytical grade reagents. Prepare and store all reagents at room temperature (unless indicated otherwise). Diligently follow all waste disposal regulations when disposing of waste materials. We do not add sodium azide to the reagents.

1. 6-well cell culture plates.

2. Complete culture medium for HEK293 cells: Dulbecco's Modified Eagle's medium F12 (DMEM–F12 1:1), supplemented with 10 % fetal bovine serum (FBS) and 1 %-glutamine.

3. Lipofectamine 2000 DNA Transfection Reagent (Life Technologies). Other transfection reagents might be also suitable if transfection efficiency is high.

4. Opti-MEM® medium (Life Technologies).

5. Sterile, 0.22 μm filtered 0.01 % poly-l-lysine solution.

6. Sterile Dulbecco's phosphate-buffered saline (D-PBS).

7. Stimulation Buffer: 147 mM NaCl, 10 mM HEPES, 13 mM d-(+)-glucose, 2 mM KCl, 2 mM $CaCl_2$, and 1 mM $MgCl_2$, pH 7.4. After preparation sterilize the buffer through 0.22 μm filters (*see* **Note 1**).

8. Nigericin sodium salt from *Streptomyces hygroscopicus* (*see* **Note 2**).

9. Coelenterazine-h (*see* **Note 3**).

*2.2 Materials
and Equipment*

1. Cell culture CO_2 incubator set at 5 % CO_2 and 37 °C.

2. Biological safety cabinets, Class II.

3. A plate reader for luminescence recording equipped with two emission filters close to 480 nm and 535 nm able to sequentially detect light at the two distinct wavelengths, including temperature control and coupled to reagent injectors, such as the Synergy™ Mx from BioTek Instruments.

4. A plate reader for fluorescence measurement able to excite close to 480 nm and read emission close to 535 nm. It could be the same plate reader than the one used for luminescence detection.

5. 96-well plate, white, flat and micro-clear bottom, cell culture treated, sterile with lids for BRET measurements (*see* **Note 4**).

6. 96-well plate, black, flat and micro-clear bottom, cell culture treated, sterile with lids for fluorescence detection.

3   Methods

Carry out all procedures with sterile and pyrogen-free material in biological safety cabinets, Class II at room temperature unless otherwise specified.

*3.1 Expression
of NLRP3 BRET
Sensors in HEK293
Cells*

1. Seed HEK293 cells in 6-well plates using 2 ml of complete culture medium at a density of 1.4×10^6 cells per well and incubate at 37 °C and 5 % CO_2 for 24 h until transfection.

2. Transfect cells following manufacturer's instructions. Briefly prepare a mix of 1 μg of total DNA in 50 μl Opti-MEM (tube A) and a mix of 3 μl of Lipofectamine 2000 diluted in 50 μl

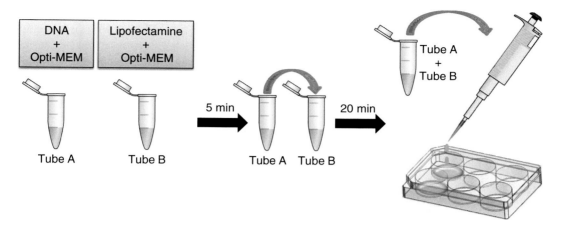

Fig. 3 Representative scheme of HEK293 cells transfection protocol using Lipofectamine 2000 reagent

Opti-MEM (tube B). Incubate tubes 5 min at room temperature and then add the volume of tube A into tube B and gently mixing. Incubate the mixture 20 min at room temperature. In the meantime, remove cell culture medium from cells and add 800 µl of fresh warm complete medium (keep the cells at 37 °C and 5 % CO_2 during the rest of the time). Then add the Lipofectamine–DNA mixture to the cells (drop to drop) and gently swirl to mix (Fig. 3). Incubate at 37 °C and 5 % CO_2 for 24 h (*see* **Notes 5** and **6**).

3. The following day, prepare poly-l-lysine solution to coat 96-well white plates by diluting 0.01 % poly-l-lysine solution 1:100 in D-PBS, mix well, and add 100 µl to each well. Incubate 15 min at room temperature.

4. During the incubation time detach transfected cells with 1 ml of complete medium by gentle pipetting up and down (**Note 7**) and transfer them into a sterile tube containing 1 ml of complete medium and carefully mix.

5. Aspirate poly-l-lysine solution from each well, wash wells twice with D-PBS and dispense 100 µl of cell suspension per well. Incubate the plate at 37 °C and 5 % of CO_2 for 24 h (*see* **Note 8**).

6. In addition, in both intermolecular and intramolecular interaction studies, in a black 96-well plate seed duplicate wells with either cells transfected with donor alone or cells transfected with the constructions fused to the BRET donor and acceptor (*see* **Note 9**).

3.2 Determination of Specific Intermolecular (Model-A) or Intramolecular (Model-B) BRET Signal

1. Program the microplate reader according to manufacturer's instructions.

2. Using the black plate (*see* **Note 9**), wash the cells two times with PBS and read the YFP fluorescence after laser excitation of the protein, to determine expression of the BRET acceptor construction.

3. To record BRET signal, wash wells from the white plate (*see* **Note 9**) once with stimulation buffer (be careful not to detach cells) and dispense 40 μl per well of stimulation buffer followed by 10 μl well of 30 μM coelenterazine-h prepared also in stimulation buffer. Final concentration of coelenterazine-h will be 5 μM (*see* **Note 10**).

4. Place the plate into the reader and start reading immediately (*see* **Note 11**).

5. BRET value, expressed in milliBRET units (or mBU), can be determined with the following equation:

$$\text{BRET}(\text{mBU}) = \left(\left(\frac{\text{Lum}(535\text{nm})}{\text{Lum}(480\text{nm})} \right)^{\text{donor+acceptor}} - \left(\frac{\text{Lum}(535\text{nm})}{\text{Lum}(480\text{nm})} \right)^{\text{rLuc } only} \right) \times 1000$$

6. In case of studying intermolecular BRET (model-A), for each condition of transfection, determine the ratio acceptor/donor by dividing the fluorescence obtained with the black plate by the luminescence measured at 485 nm in the white plate during BRET recording. Report these values (X axis) to the BRET value (Y axis) to get a saturation curve. In case of studying intramolecular BRET (model-B), report the value of fluorescence (X axis) to the BRET value (Y axis).

7. Determine the BRET specificity according to Fig. 2.

3.3 Detection of NLRP3 BRET Signal Variation During Activation

1. Program the microplate reader according to manufacturer's instructions (injection timing, injection speed, delivery volume, temperature), clean the injectors with distilled water, prepare the stimulus in the hood, and fill up the injector with the stimulus (*see* **Note 12**).

2. To prepare each BRET experiment (one row each time, Fig. 4), wash wells from the white plate once with stimulation buffer (be careful not to detach cells) and dispense 40 μl per well of stimulation buffer followed by 10 μl well of 30 μM coelenterazine-h prepared also in stimulation buffer. Final concentration of coelenterazine-h after stimulation injection will be 5 μM (*see* **Note 10**).

3. Place the plate into the reader and start reading immediately (*see* **Note 11**). After 5 min of basal BRET signal reading, inject 10 μl of stimulus per well (*see* **Note 13**). Luminescence recording is measured at 480 nm and 535 nm every 10 s for 20 min. The measures before stimulation represent the basal level of BRET signal within each experiment and will be used to analyze the BRET response during inflammasome activation.

4. Add NLRP3 inflammasome activator. We classically used the selective K⁺ ionophore nigericin (*see* **Note 14**).

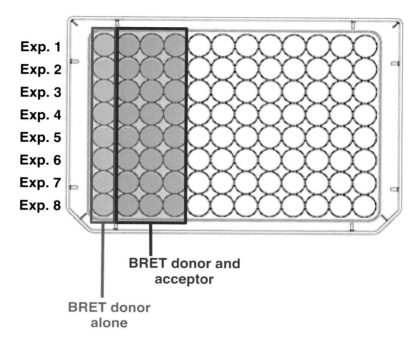

Fig. 4 Plate preparation scheme. Transfected cells with the BRET donor–acceptor construct/s or the donor alone construct are seeded in a white 96-well plate as indicated. Four wells were prepared for each experiment or experimental condition (Exp. 1–8), one with the BRET donor transfected cells and three wells with the BRET donor and acceptor combination to perform triplicate recordings

5. Determine BRET value (*see* above Subheading 3.2, **step 5**)

6. Express the mBU values as the percentage of BRET signal relative to the baseline, been this basal BRET signal the average of the mBU values from the 5 min before stimulus injection (*see* **Note 15**). Then calculate the Mean from the three final % mBU values, the Standard Deviation and the Standard Error of Mean (SEM).

7. Represent in a dispersion plot with the normalized BRET signal values calculated (Y axis) against time (X axis) to produce a kinetic profile (Fig. 5).

4   Notes

1. For long storage periods keep the buffer at 4 °C.

2. Prepare nigericin stock solution at 10 mM in ethanol and store at –20 °C.

3. Coelenterazines are poorly soluble in water and must be resuspended in ethanol or methanol preparing a 1 mM stock solution. Keep stock solution at –20 °C and protect it from light.

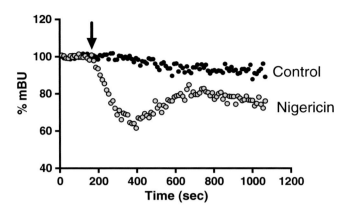

Fig. 5 Representative plot of intramolecular NLRP3 BRET signal (model-B) kinetic in response to vehicle (*black circles*) or 5 μM nigericin as a NLRP3 inflammasome activator (*grey circles*). The *arrow* indicates when nigericin was added

For prolonged stability, resuspend coelenterazine in acidified ethanol (10 ml of ethanol and 200 μl of 3 N HCl) and store at −80 °C.

4. 96-well plate with white bottom could be used but will not allow to visualize cells during the different steps of the experiment. If using micro clear bottom plates, a white backing tape can be stuck on the bottom of the 96-well plate the day of recording and will increase luminescence signal by reflection. Note that BRET recording using a plate reader can be performed on other format plate (i.e., 384-well plates) depending on the assay and/or luminescence signal strength.

5. It is important to perform two sets of transfections. One cell transfection with the BRET donor construct alone, which will be used to monitor the signal at 530 nm from the donor, and one cell transfection with the donor–acceptor constructs to measure the BRET signal.

6. To determine the specificity of intermolecular BRET signal (model-A, Fig. 2a), perform BRET saturation curve by transfecting a constant amount of NLRP3-rLuc donor with increasing concentrations of NLRP3-YFP acceptor. In case of specific interaction, BRET signal will increase and become saturable with increasing amounts of acceptor. On the other hand, in case of random collision between proteins, BRET signal will be low and will increase linearly with higher quantity of BRET acceptor. To analyze if BRET recorded from NLRP3 in model-B is intramolecular or intermolecular as represented in Fig. 2b, transfect increasing concentrations of NLRP3 tagged with rLuc and YFP. Intramolecular BRET signal is not dependent on expression level of the protein.

7. HEK293 cells are easily detached without the use of trypsin; however, similar results can be obtained by detaching the cells with trypsin.

8. For each BRET experiment prepare one well with cells transfected with donor alone and three wells with the cells transfected with the constructions donor and acceptor, prepare as many rows as different experimental conditions you want to test (Fig. 4).

9. The white plate is for dual luminescence detection at 480 nm and 535 nm, and the black plate is for YFP fluorescence reading.

10. The dilution of coelenterazine-h in stimulation buffer has to be performed immediately before using it and solution has to be kept in dark conditions. It is important to take into account when prepare the diluted coelenterazine-h that the final volume in the well will be 60 μl after cell stimulation.

11. After coelenterazine-h addition wait around 6 min to allow BRET signal stabilization.

12. If using a plate reader with a single injector, the injector has to be purged with distilled water at least twice every time a different stimulation is tested. Prepare enough volume of solution to fill up the injector and to inject in the wells. Take into account when preparing the stimulation solution that the final volume in the well after injection will be 60 μl, therefore prepare the right concentrated solution.

13. Use a slow injection speed to decrease mechanically stimulation of the cells. We normally use 225 μl/s.

14. Prepare enough volume of solution when injecting to fill up the injectors and to dispense into the wells. In the BioTek Synergy Mx plate reader we require 1.1 ml to fill up the injectors, so it is recommended to prepare at least 1.5 ml of 60 μM nigericin solution. When preparing nigericin dilution, take into account that the injection volume will be 10 μl.

15. To calculate the basal BRET signal average, discard the first data points until the luminescence is stabilized.

Acknowledgements

This work was supported by the European Research Council and *instituto salud carlos III*-FEDER grants to P.P. F.M.-S. was supported by a *Sara Borrell* fellowship from the *instituto salud carlos III*. V.C. was supported by the *Institut National de la Santé et de la Recherche Médicale*.

References

1. Pfleger KDG, Eidne KA (2006) Illuminating insights into protein-protein interactions using bioluminescence resonance energy transfer (BRET). Nat Methods 3:165–174

2. Pfleger KDG, Seeber RM, Eidne KA (2006) Bioluminescence resonance energy transfer (BRET) for the real-time detection of protein-protein interactions. Nat Protoc 1:337–345

3. Xu Y, Kanauchi A, von Arnim AG, Piston DW, Johnson CH (2003) Bioluminescence resonance energy transfer: monitoring protein-protein interactions in living cells. Methods Enzymol 360:289–301

4. Wu PG, Brand L (1994) Resonance energy transfer: methods and applications. Anal Biochem 218:1–13

5. Siddiqui S, Cong W-N, Daimon CM, Martin B, Maudsley S (2013) BRET biosensor analysis of receptor tyrosine kinase functionality. Front Endocrinol (Lausanne) 4:46

6. Xu Y, Piston DW, Johnson CH (1999) A bioluminescence resonance energy transfer (BRET) system: application to interacting circadian clock proteins. Proc Natl Acad Sci U S A 96:151–156

7. Compan V, Baroja-Mazo A, López-Castejón G, Gomez AI, Martínez CM, Angosto D et al (2012) Cell volume regulation modulates NLRP3 inflammasome activation. Immunity 37:487–500

Chapter 11

Measuring NLR Oligomerization IV: Using Förster Resonance Energy Transfer (FRET)-Fluorescence Lifetime Imaging Microscopy (FLIM) to Determine the Close Proximity of Inflammasome Components

Catrin Youssif, Bárbara Flix, Olivia Belbin, and Mònica Comalada

Abstract

Intracellular signaling and cellular activation have been demonstrated to reside on multi-protein complexes rather than in isolated proteins. Consequently, techniques to resolve these complexes have gained much attention over the last few years. Förster Resonance Energy Transfer (FRET) coupled with Fluorescence Lifetime Imaging Microscopy (FLIM) is a powerful tool to discriminate direct interactions between two proteins within a multi-protein complex. Here, we present the use of FRET-FLIM as an experimental tool for the interpretation of the inflammasome composition. We also introduce some considerations required for the correct use of this technique and the control experiments that should be implemented.

Key words FRET, FLIM, Inflammasome, Immunofluorescence, Macrophages, Endogenous proteins, NLRP3, Caspase-1, Multi-protein complex

1 Introduction

Inflammasomes are cytosolic multi-protein complexes involved in bacterial recognition and inflammation, playing a critical role on the secretion of IL-1 related cytokines and pyroptosis [1]. Inflammasomes assemble around a set of core components that include (1) a sensor protein, usually a member of the Nucleotide-binding and oligomerization domain-like receptor (NLR) or Absent in melanoma 2 (AIM2) families (2), an adaptor protein, mainly the Apoptosis-associated speck-like protein (ASC), and (3) an inflammatory caspase (CASP), such as CASP-1. The inflammasome activity, type, and complex composition depends on the nature of the particular stimulus. The list of agents that are able to activate inflammasomes is constantly increasing and includes pathogen-derived proteins, lipids, nucleic acids, and polysaccharides, as well as high levels of ATP and alterations in cell osmolarity

Francesco Di Virgilio and Pablo Pelegrín (eds.), *NLR Proteins: Methods and Protocols*, Methods in Molecular Biology, vol. 1417, DOI 10.1007/978-1-4939-3566-6_11, © Springer Science+Business Media New York 2016

[1, 2]. It is therefore not surprising that to correctly sensor all these stimuli, a plethora of proteins such as NLRP3, NLRP6, NLRC3, NLRC4, AIM2, and NLR-family apoptosis inhibitory proteins (NAIPs) have been involved in inflammasome activation. Moreover, recently, several so-called unconventional inflammasomes have emerged, in which ASC is not required or where caspases other than CASP-1, such as CASP-4, -5, -8, or -11 are involved [1–4]. The diversity of inflammasome components and responses has been also extended by cross talk with other Pattern Recognition Receptors (PRR) and immune signaling events [3, 4], or by interaction of the NLR family member with other inflammasome assembly initiator proteins such as Thioredoxin-interacting protein (TXNIP) or NAIP proteins [4].

Here, we describe an approach to analyze the interaction of endogenous inflammasome components in mouse Bone Marrow-Derived Macrophages (BMDM). BMDM are primary cultures that possess unmodified activation machinery in response to physiological stimuli. Based on that, they constitute an excellent model to study the inflammasome response under physiological conditions.

The combined use of Förster Resonance Energy Transfer (FRET) and Fluorescence Lifetime Imaging Microscopy (FLIM) is an established and robust method to demonstrate close proximity (<10 nm) between two proteins with on a nanosecond timescale [5, 6]. Conceptually, FRET-FLIM is based on the phenomenon of FRET [7], in the lifetime of the donor fluorophore is shortened in the presence of an acceptor fluorophore at a maximal distance of 10 nm. Hence, by immunostaining two proteins with fluorophores at different emission wavelengths, the fluorescence lifetime of the donor fluorophores, as detected by FLIM [8, 9], is inversely proportional to the distance between the two-labeled proteins. A shortened lifetime in the presence versus absence of the acceptor fluorophore is therefore indicative for a distance <10 nm between the two proteins, a resolution that is currently not reached by using conventional microscopy techniques (Fig. 1).

One of the main advantages of FRET-FLIM compared to other techniques to detect protein–protein interactions (e.g., co-immunoprecipitation), is the fact that it is a direct measurement of the distance between two fluorophores. Moreover, in contrast to the intensity-based FRET technique, FRET-FLIM measurements are insensitive to the concentration of fluorophores and can thus filter out artifacts introduced by variations in the concentration and emission intensity across the sample and as such, can avoid false positive signals. Thus, the combination of FLIM with FRET increases the sensitivity compared to the appliance of using FRET alone [10, 11]. Furthermore, FRET-FLIM has been used to demonstrate direct protein–protein interactions in both, fixed and live cellular sections [12] using both fluorescently tagged [13] and immunostained proteins [14] and can be used for both endogenously expressed or

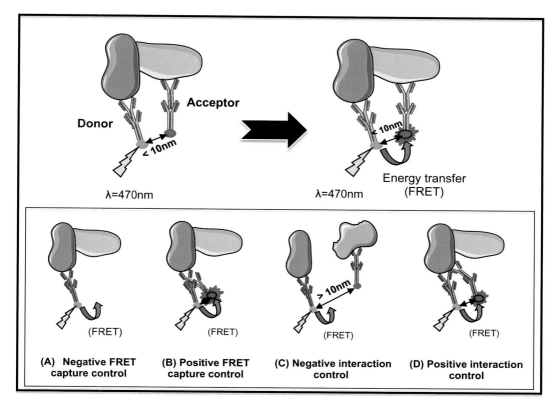

Fig. 1 Schematic representation of the basic principle of FRET-FLIM technique

transfected proteins. Therefore, the FRET-FLIM technique is a good candidate to explore the composition of inflammasomes and to detect endogenously expressed inflammasome proteins in fixed myeloid cells, thereby avoiding artifacts caused through over-expression of proteins by transfection. The sensitivity of the technique allows us to detect interactions even when the levels of endogenous proteins of interest are low and/or the proteins of interest are unequally expressed. However, the distance added by antibodies used for immune-detection has to be taken in consideration, since this could lead to lower FRET efficiency.

The high sensitivity of FRET-FLIM requires adequate robust controls for FRET capture. The most common controls used to establish and validate the FRET-FLIM assay (Fig. 1) are controls for FRET capture (A) the donor lifetime measured in the absence of the acceptor (negative FRET capture control) and (B) the donor lifetime in the presence of an acceptor fluorophore at a distance <10 nm (positive FRET capture control). This can be achieved either using two primary antibodies raised in different species against the same protein or by using two secondary antibodies against the same primary antibody. Additionally, controls for protein–protein interactions should also be used: (C) two proteins predicted to colocalize but not interact (negative interaction control) and (D) two proteins known to interact with each other in the cellular system being studied. Since the emission wavelength of the donor fluorophore must overlap with the excitation wavelength of the acceptor fluorophore, the 488 nm and 555 nm Alexa-tagged antibodies are commonly used as the donor and acceptor, respectively, however, there are other fluorophore pairs other fluorophore pairs such as GFP and RFP or yellow (venus) and cyan (cerulean) can also be used for FLIM.

For optimal detection of inflammasome components, FRET capture can be achieved as previously described [13, 14]. Briefly, the donor fluorophore is excited using a 470 nm pulsed laser with 40 MHz repetition rate. The fluorophores are subsequently detected using a time-correlated single photon counting detector. A fluorescence filter (500–550 nm) limits the detection to that of the donor fluorescence only (Alexa Fluor 488). SymphoTime v5.2 software (PicoQuant) is used to measure fluorescence lifetimes on a pixel-by-pixel basis with high spatial resolution. All confocal pictures are acquired with a Laser Scanning Confocal Inverted Microscope Leica TCS SP5-AOBS, equipped with PicoQuant LSM Upgrade Kit for FLIM and FCS, which is used to perform FRET-FLIM experiments. The captured lifetime of the donor fluorophore (Alexa Fluor 488) in the absence of the acceptor fluorophore (Alexa Fluor 555) is measured using the negative FRET capture controls. It must be noted, that while it is common in biological systems for a fluorophore to emit two lifetimes (L1 and L2), in some systems a decay curve with a single lifetime (L1) may be more appropriate. The goodness-of-fit of a 2 versus 1-exponential model should therefore be assessed using the χ^2 value as calculated by Symphotime (choosing the model that gives a value closest to 1). The mean donor lifetime (L1 for 1-exponential or L1 and L2 for 2-exponentials) in the absence of an acceptor is calculated for at least ten individually captured negative FRET capture control cells. Subsequently, the donor lifetime in the presence of the acceptor fluorophore (Alexa Fluor 555) is calculated for at least 10 individually captured cells for each of the remaining control and experimental conditions. This is achieved by fixing the mean donor lifetime(s) calculated using the negative FRET capture controls and adding an extra exponential to the model (L2 for 1-exponential or L3 for 2-exponentials), which represents the donor–acceptor lifetime. The average lifetime of the exponentials (τInt) can be calculated as the average amount of time for which the fluorophores remain in their excited state after the onset of excitation and therefore increases with intermolecular distance. The FRET efficiency, or EFRET, is calculated as the proportion of the donor molecules that have transferred excitation state energy to the acceptor molecules, such that the EFRET increases with decreasing intermolecular distance. Additionally, false-color images can be produced by using a pseudo-color scale based on the τInt value of each pixel, on a pixel-by-pixel basis, over the entire image (Fig. 2). All experiments should be performed in triplicate providing a minimum of 30 different FLIM captures per condition across three independent experiments. Each experiment should be analyzed independently, in order to control for inter-experimental variation of factors, such as scattered light. However, using the EFRET values and a fixed scattered light we can compare different experiments (Fig. 2). As an illustrative example, the EFRET and pseudo-color images of the NLRP3 and CASP-1 interaction in BMDM, non-stimulated and stimulated, with LPS plus ATP or nigericin are shown (Fig. 2), demonstrating that the NLRP3 inflammasome is assembled in BMDM only in the presence of its activating stimuli.

In conclusion, the use of FRET-FLIM supported with adequate control experiments can be a powerful tool for the confirmation of the complex composition of inflammasomes and their cellular response.

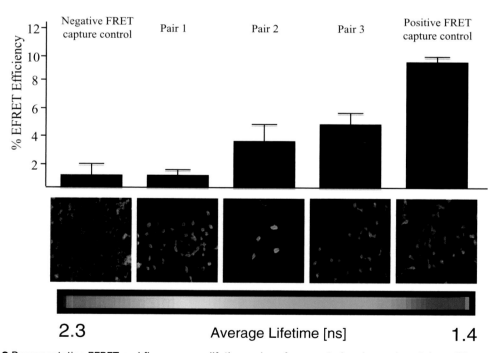

Fig. 2 Representative EFRET and fluorescence lifetime values from control and experimental conditions using the FRET-FLIM technique. Negative FRET capture control: No acceptor (only mCASP-1); Positive FRET capture control (mCASP-1 recognized by 2 secondary antibodies); Pair 1 (CASP-1-NLRP3 in absence of macrophage stimuli); Pair 2 (CASP-1-NLRP3 in LPS + ATP treated macrophages); Pair 3 (CASP-1-NLRP3 in LPS + Nigericin treated macrophages)

2 Materials

2.1 Bone Marrow-Derived Macrophages (BMDM)

1. Class II cell culture hood.
2. Incubator maintained at 37 °C and 5 % CO_2, 95 % air.
3. Sterile surgical material (forceps and scissors).
4. 25 gauge needle.
5. 5 mL Syringe.
6. Cell scrapers.
7. Petri dishes (150 mm).
8. Complete Dulbecco's modified Eagle's medium (DMEM) supplemented with 30 % L-cell conditioned medium as a source of M-CSF (Macrophage-colony stimulating factor) (*see* **Note 1**), 20 % of heat-inactivated fetal bovine serum (FBS, inactivated at 56 °C for 30 min, store aliquots at –20 °C), 2 mM L-glutamine, and 1 % of 10,000 U/mL penicillin/10 mg/mL streptomycin (P/S).
9. Ultra Pure Lipopolysaccharide from Escherichia coli 0111:B4 (LPS).
10. Nigericin, as an inflammasome activator activator.
11. Adenosine 5′-triphosphate disodium salt hydrate (ATP), as inflammasome activator.

2.2 Immunofluorescence

1. 8-well glass chamber slides, such as the PEZGS0816 (Millipore).
2. Phosphate buffered saline solution (PBS)-1× (500 mL): 4 g NaCl, 0.1 g KCl, 0.38 g $Na_2HPO_4 \times 2H_2O$, 0.1 g KH_2PO_4 completed with ddH_2O to reach a volume of 500 mL.
3. 100 % methanol, aliquots maintained at –20 °C.
4. Triton X-100.
5. Normal donkey serum aliquots maintained at ≤20 °C.
6. Anti-CD16/CD32 (Fcγ III/II R,) Fc receptor blocking antibody.
7. Primary antibodies against the proteins of interest. *See* Table 1.
8. Secondary antibodies conjugated to Alexa Fluor 488® for donor and Alexa Fluor 555® for acceptor. *See* Table 1.
9. Blocking buffer Stock solution (2 mL): 0.1 mL normal donkey serum (dilution 1/20), 0.2 μL Triton X-100 (final concentration 0.01 %), and 1.9 mL PBS-1×. Prepare fresh solution daily (*see* **Note 2**).
10. Blocking buffer with CD16/CD32 blocking antibody: add 20 μL CD36/CD32 Blocking antibody (dilution 1/50) to 1 mL of blocking buffer stock solution.

Table 1
Primary and secondary antibodies used for FRET-FLIM and immunofluorescence techniques

Antibody	Host	Reference	Provider	Suggested dilution
NLRP3	Rabbit	Ab91525	Abcam	1/100
NLRP3	Rabbit	AP-22017PU-N	Acris	1/125
NLRC4	Rabbit	AP-21931SU-S	Acris	1/50
CASPASE-1	Mouse	Sc-56036	Santa Cruz	1/50
ASC	Rabbit	AP-06792PU-N	Acris	1/100
Anti-mouse Alexa Fluor 488	Donkey	A21202	Invitrogen	1/200
Anti-rabbit Alexa Fluor 488	Donkey	A21206	Invitrogen	1/200
Anti-rabbit Alexa Fluor 555	Goat	A31572	Invitrogen	1/200
Anti-goat Alexa Fluor 555	Goat	A21432	Invitrogen	1/200

11. Incubation buffer (2 mL): 0.1 mL normal donkey serum and 1.9 mL PBS-1×. Prepare daily fresh solution.

12. Mowiol mounting medium for immunofluorescence staining.

13. Conventional Fluorescence Microscope with appropriate filter sets is recommended in order to check the quality and intensity of the immunofluorescence staining prior to FRET-FLIM analysis.

2.3 Image Capture, FLIM Measurement and Analysis

1. Laser Scanning Confocal Inverted Microscope, such as the Leica TCS SP5-AOBS equipped with PiccoQuant LSM Upgrade kit for FLIM and FCS (Leica) and the SymPhoTime (2011) software. Other similar microscopes are also suitable.

3 Methods

3.1 Bone Marrow-Derived Macrophages [15]

1. Prepare sterile surgical materials, needles and syringes.

2. Kill mice by cervical dislocation. Remove the legs and collect the femur and tibiae. Working in sterile conditions in a laminar flow hood, cut the end of the bones and flush them with DMEM using a syringe with an adequate injection needle.

3. Disaggregate the bone marrow by pipetting up and down. Culture the cells in complete DMEM medium. The bone marrow isolated from one mouse is usually divided into four petri dishes (40 mL/plate).

4. To obtain an almost homogenous population of macrophages, culture the cells for 6 days at 37 °C in 5 % CO_2 without changing the media.

Table 2
Example of a FRET-FLIM experiment optimization with suggested controls

FLIM negative capture control			
Stimuli	None		
Donor 1:	**mCASP-1**	Donor 2:	**DAM-488**
Acceptor 1:	**None**	Acceptor 2:	**None**
FLIM negative interaction control			
Stimuli	None		
Donor 1:	**mCASP-1**	Donor 2:	**DAM-488**
Acceptor 1:	**rArginase-1**	Acceptor 2:	**DAR-555**
FLIM positive capture control 1			
Stimuli	None		
Donor 1:	**mCASP-1**	Donor 2:	**DAM-488**
Acceptor 1:	**rCASP-1**	Acceptor 2:	**DAR-555**
FLIM positive capture control 2			
Stimuli	None		
Donor 1:	**mCASP-1**	Donor 2:	**DAM-488**
Acceptor 1:	**None**	Acceptor 2:	**GAD-555**
Interaction of inflammasome components			
Stimuli	LPS (1 μg/mL) for 18 h and last 30 min add ATP (5 mM)		
Donor 1:	**mCASP-1**	Donor 2:	**DAM-488**
Acceptor 1:	**rNLRP3**	Acceptor 2:	**DAR-555**

DAM Donkey anti-Mouse, *DAR* Donkey anti-Rabbit, *GAD* Goat anti-Donkey, *m* mouse, *r* rabbit, *Arginase-1* a cytosolic protein that is not known to interact with inflammasome components

5. After 6 days, detach the adherent cells by scraping in the same media, centrifuge at $500 \times g$ at 4 °C for 5 min.

6. Count the cells using a Neubauer cell chamber or any other equivalent methodology.

7. Seed 50,000 macrophages/well in 8-well glass chamber slides in a total volume of 500 μL of complete media and allow cells to adhere (>30 min) prior to stimulation.

8. Treat the cells with the desired stimuli for the established times and corresponding controls. *See* some examples in Table 2 (*see* **Notes 3** and **4**).

3.2 Immunofluorescence

1. Remove the culture media and immediately fix the cells with 500 μL ice-cold 100 % methanol for 5 min at room temperature (RT) (*see* **Note 5**).

2. Carefully aspirate the methanol and wash 3 times with PBS-1× at RT.

3. Aspirate PBS-1× and incubate the samples with 100 μL of blocking buffer containing CD16/32 blocking antibody at RT with shaking during 45 min (*see* **Note 6**).

4. Carefully discard the blocking buffer containing CD16/32 blocking antibody and add 100 μL of blocking buffer (without CD16/32 antibody), containing the primary antibodies for the donor and the acceptor at established concentrations (*see* Table 1) for 2 h at RT with shaking (*see* **Notes 7** and **8**).

5. Wash the cells 3 times for 10 min (with shaking) in PBS-1× at RT. Cover the plate with aluminum foil to avoid light exposure.

6. Aspirate the wash solution and add 100 μL of incubation buffer with 1/200 dilution of the secondary antibody (Alexa Fluor 488) for 1 h at RT with shaking (*see* **Notes 9** and **10**).

7. Remove the incubation buffer and wash the cells 3 times for 10 min (with shaking) in PBS-1× at RT.

8. Add 100 μL of incubation buffer containing a 1/200 dilution of the secondary antibody (Alexa Fluor 555) for 1 h at RT with shaking (*see* **Notes 9** and **10**).

9. Wash the cells 3 times for 10 min (with shaking) in PBS-1× at RT.

10. Remove the PBS-1× and mount slides carefully with three drops of Mowiol and a cover glass.

11. Keep the slide at 4 °C until use (*see* **Notes 11** and **12**).

3.3 Image Capture

1. Use a Laser Scanning Confocal Inverted Microscope Leica TCS SP5-AOBS, equipped with PiccoQuant LSM Upgrade Kit for FLIM and FCS. The SymPhoTime software will be used to measure fluorescence lifetimes on a pixel-by-pixel basis with high spatial resolution (*see* **Note 13**).

2. Turning on the PC, microscope and lasers: Start LAS and Symphotime software; turn on the fluorescence lamp and turn on the detector channel to 1 (turn laser dial to 0) (*see* **Notes 14** and **15**).

3. Introduce the 8-well glass chamber slides in the Laser Scanning Confocal Inverted Microscope Leica TCS SP5-AOBS, equipped with PiccoQuant LSM Upgrade Kit for FLIM and FCS.

4. Place oil on objective (63× oil).

5. Focus cells using eyepiece.

6. Ensure the 488-561 nm laser filter is in place.

7. Setting up a FLIM experiment (*see* **Note 16**).

8. In the LASAF software (bottom screen): Select SMD FLIM WIZARD from the Leica System drop-down list (top left of the screen).

9. Setup Imaging Tab:

   (a) Select xyz

   (b) 512 × 512 pixel images

   (c) 400 Hz

   (d) In Channel 1 (PMT2), select Leica/EFFP from drop-down list and set the correct wavelength on the scale for 488 capture

   (e) Set the 488 laser capture to 20 %

   (f) In channel 2 (PMT3), select Leica/TRITC from drop-down list and set the correct wavelength on the scale for 555 capture

   (g) Set the 555 capture to 20 %

   (h) MFP > substrate

   (i) X1-port setting > mirror, 63× objective oil

10. Setup FLIM Tab:

    (a) 256 × 256 pixels

    (b) 100 Hz

    (c) Deselect PMT2, PMT3 and visible light channel

    (d) Click Ext Pulsed

    (e) Set the 470 laser to 1

    (f) PMTAPD1 > Active > Scan-BF

    (g) MFP > DD470/640

    (h) X1-port settings > •---

11. Measurements Tab:

    (a) Acquire > acquire until max 1000 photons/pixel

    (b) MFP > DD470/640

    (c) X1-port settings > •---

12. Setup Imaging Tab: New experiment.

3.4 FLIM Measurements (see Notes 17 and 18)

1. Place the slides on microscope and focus a cell with the eyepiece.

2. In the LASAF software setup Imaging Tab:

   (a) New experiment

   (b) Click Live

   (c) Focus image > z-position, gain and offset

   (d) Click start to take photo

   (e) Save images

3. In the LASAF software setup FLIM Tab:

   (a) Run FLIM test

   (b) In the popup window look at the Peak Count Rate

   (c) Adjust to approx. 1000 counts by adjusting the micrometer on top of the laser box (*see* **Note 19**)

   (d) Click Stop on the left screen (bottom right)

4. In the LASAF software setup measurement Tab: Run FLIM.

5. In the SymPhoTime500 software a new file will appear called Base Membrane 1, rename the file and continue for all conditions.

3.5 Analysis FLIM Results

1. In the SymPhoTime500 software go to File > Load workspace or choose from list below it.

2. Select the first cell from BaseName folder.

3. Duplicate file for analysis (Edit > Duplicate).

4. On popup window to the right click the 3rd button (FASTFLIM).

5. Click Fitting and TCSPC and a graph will open.

6. Move the pink lines to the start and end of the straight line (*see* **Note 20**).

7. Click Recalculate.

8. In fitting options click Use MLE.

9. On the right, click Clear.

10. Click FIT.

11. Make a note of the τamp value.

12. Compare using 1 or 2 components (χ value should be roughly 1 (preferably between 0.8 and 1.2)) (*see* **Note 21**).

13. Look for small residuals that randomly cross zero (without a trend above or below the line) and Click Export.

14. Repeat for all the Negative FRET capture controls used in the experiment.

15. Open all of the ".dat" files for the negative controls and take down the L1 value.

16. Calculate the mean lifetime (L1) (and L2 for a 2-exponential model) for the negative FRET capture controls.

17. Duplicate the first positive control folder and open.

18. Repeat **steps 6–9**.

19. Add the donor–acceptor lifetime (*see* **Note 22**):

   (a) If the negative controls were best fit to 1-component model: Select 2 components. In L1 type the mean L1 value for the negative controls. To the right of L1 click on *?srch* twice so that it says: !Fixed

(b) If the negative controls were best fit to 2-component model: Select 3 components. In L1 and L2 type the mean L1 and L2 values for the negative controls, respectively. To the right of L1 and L2 click on *?srch* twice so that it says !Fixed

20. click FIT.

21. Make a note of the τamp value.

22. Click Export.

23. Repeat for all the Positive Controls and Experimental conditions.

3.6 Image Analysis

To export the image files:

1. Open and duplicate the file for the first cell.

2. Click Colors and Lifetime Histogram.

3. On the right-hand panel play with the limits of the graph (typically 1.5–2.2) comparing the positive and negative controls until you see a difference in color that you sufficiently differentiate between the positive and negative controls. These limit values should be used for ALL cells analyzed.

4. Click Enter to save the values and update the graph.

5. Click Export BMP at the top of the screen (exports photo of cell).

6. Right click on graph and export as BMP (exports graph image).

7. Click Export graphs (exports numbers for graph).

To calculate the FRET energy transfer efficiency use the following equation:

$$\text{EFRET} = \frac{(\text{Mean τamp negative controls} - \text{Mean τamp condition})}{\text{Mean τamp negative controls}} \times 100$$

4   Notes

1. L-cell conditioned medium is obtained from supernatants of mouse L929 fibroblast cell line.

2. Donkey or goat serums are commonly used for blocking depending on the origin of secondary antibody.

3. It is recommended to use an adequate number of controls, at least, two negative controls and two positive controls for FLIM. However, since FRET-FLIM capture must be performed on the same day for all conditions within an experiment, for time constraints, experiment will include four experimental conditions.

4. The primary antibodies must be raised in different species for differential recognition by the secondary antibodies.

5. Alternatively, the 8-well glass chamber slides with methanol can be stored at −20 °C for up to several weeks.

6. CD16/CD32 blocking Fc Antibodies are used to eliminate the majority of unspecific antibody capture by BMDM, and therefore background signal.

7. For a stronger and more specific immunostaining signal, instead of incubating for 2 h at RT, the samples can be incubated O/N at 4 °C (with shaking) in the blocking buffer containing the primary antibodies.

8. It is recommended to optimize the dilutions of primary antibodies by conventional immunofluorescence technique prior to FRET-FLIM analysis, since some variability can be observed between commercial providers, lot number and storage conditions of the antibodies. A high, specific signal is required for adequate FRET capture.

9. It is recommended to centrifuge the primary and secondary antibodies at $10,000 \times g$ for 10 min before dilution to remove precipitates.

10. The dilution should be validated depending on the commercial provider of the secondary antibodies.

11. Prior to the FRET-FLIM capture, it is recommended to first check the antibodies to be used as donor and acceptors and the immunostaining signal of each slide using a confocal or fluorescence microscope in order to avoid a waste of time.

12. For FRET-FLIM analysis, samples should not be stored longer than 3 days.

13. From each experimental condition, ten focal field images containing ~15–60 cells will be captured. It is advisable to capture individual cells rather than a population of cells and to avoid background interference. To establish and validate the FRET-FLIM capture, Alexa Fluor 488 lifetime decay is measured in the absence of the acceptor as negative FRET capture control.

14. It is recommended to consult the expertise of an experienced microscopy user when performing FLIM capture in order to ensure that excitation and capture settings are appropriately used.

15. The following settings are for Alexa Fluor 488 and Alexa Fluor 555.

16. There is no sequential option in FLIM, so if sequential photos are required, it is necessary to exit the FLIM Wizard and capture photos in the Leica LAS software. On returning to the FLIM Wizard, all FLIM settings will need to be reapplied, which can be time consuming.

17. It is advised to start with the negative FRET capture control followed by the positive FRET capture control to ensure FRET capture.

18. Each individual FLIM capture can take approximately 5 min.

19. If a value of 1000 is not reached, it is advisable to pick a different cell to avoid excessive beaching.

20. It is advised to place the bars in the same position for all analyses.

21. All lifetime values should be positive. If not, reject the condition.

22. The goodness-of-fit of the model using 1 versus 2 components can vary depending on the fluorophore and the cell system being used. Adding more components allows for greater complexity, and can therefore result in a better fit, albeit it can also produce a more artificial model.

Acknowledgement

M.C. and O.B. contributed equally to this work. The authors thank Carmen Casal and Eva Companys for their help in developing the FRET-FLIM technique presented here. O.B is a recipient of the "Miguel Servet" Grant provided by the FEDER (European Funds for Regional Development) and the Carlos III Institute of Health (Ministry for Economy and Competitivity, Spain). C.Y. is recipient of "La Caixa" fellowship.

References

1. Strowig T, Henao-Mejia J, Elinav E et al (2012) Inflammasomes in health and disease. Nature 481:278–286

2. Lamkanfi M, Dixit VM (2012) Inflammasomes and their roles in health and disease. Annu Rev Cell Dev Biol 28:137–161

3. Broz P, Monack DM (2011) Molecular mechanisms of inflammasome activation during microbial infections. Immunol Rev 243:174–190

4. Latz E, Xiao TS, Stutz A (2013) Activation and regulation of the inflammasomes. Nat Rev Immunol 13:397–411

5. Backsai BJ, Skoch J, Hickey GA et al (2003) FRET determinations using multiphoton fluorescence lifetime imaging microscopy (FLIM) to characterize amyloid-beta plaques. J Biomed Opt 8:368–375

6. Elangovan M, Day RN, Periasamy A (2002) Nanosecond fluorescence resonance energy transfer-fluorescence lifetime imaging microscopy to localize the protein interactions in a single living cell. J Microsc 205:3–14

7. Förster T. Intermolecular energy migration and fluorescence. Ann. Phys. 1948;2:55–75. doi:10.1002/andp.19484370105

8. Gadella BM, Colenbrander B, Van Golde LM et al (1993) Boar seminal vesicles secrete arylsulfatases into seminal plasma: evidence that desulfation of seminolipid occurs only after ejaculation. Biol Reprod 48:483–489

9. Straub M, Lodemann P, Holroyd P et al (2000) Live cell imaging by multifocal multiphoton microscopy. Eur J Cell Biol 79:726–734

10. Wallrabe H, Periasamy A (2005) Imaging protein molecules using FRET and FLIM microscopy. Curr Opin Biotechnol 16:19–27

11. Becker W (2008) Fluorescence lifetime imaging—techniques and applications. J Microsc 247:119–136

12. Thomas AV, Berezovska O, Hyman BT et al (2008) Visualizing interaction of proteins relevant to Alzheimer's disease in intact cells. Methods 44:299–303

13. Badiola N, de Oliveira RM, Herrera F et al (2011) Tau enhances alpha-synuclein aggregation and toxicity in cellular models of synucleinopathy. PLoS One 6, e26609

14. Flix B, de la Torre C, Castillo J et al (2013) Dysferlin interacts with calsequestrin-1, myomesin-2 and dynein in human skeletal muscle. Int J Biochem Cell Biol 45:1927–1938

15. Comalada M, Ballester I, Bailón E et al (2006) Inhibition of pro-inflammatory markers in primary bone marrow-derived mouse macrophages by naturally occurring flavonoids: analysis of the structure-activity relationship. Biochem Pharmacol 72:1010–1021

Chapter 12

Measuring NLR Oligomerization V: In Situ Proximity Ligation Assay

Yung-Hsuan Wu and Ming-Zong Lai

Abstract

The NLRP3 inflammasome is assembled in macrophages and monocytes in response to inflammatory and danger stimuli. The atypical nature of the NLRP3 complex impedes detection of NLRP3 inflammasome formation by conventional biochemical and cell biology methods. In situ proximity ligation assay (PLA) provides an alternative method of detection, localization, and quantification of protein–protein interactions in tissue and cell samples. Two primary antibodies raised in different species detect the two proteins of interest. When the proteins are in close proximity, secondary antibodies conjugated with specific DNA probes hybridize with linking oligonucleotides to form a DNA bridge between the two proteins. Amplification of the DNA bridge then facilitates detection by microscopy using fluorescence probes. Here, we describe application of in situ PLA to detect NLRP3 inflammasome assembly in mouse bone marrow-derived macrophages and human monocyte cell line THP1.

Key words In situ proximity ligation assay, Inflammasome, NLRP3 complex, Macrophages, NLRP3-ASC-caspase-1, Protein–protein interaction, IL-1β

1 Introduction

Inflammatory stimuli and danger signals such as pathogens, microbial products, and danger molecules trigger assembly of the NACHT domain-, leucine-rich repeat-, and pyrin domain-containing protein 3 (NLRP3, also known as NALP3 or cryopyrin) inflammasome in macrophages and monocytes. The formation of the NLRP3 inflammasome involves two sequential steps triggered by distinct signals. In the first stage, signals from innate receptors induce NF-κB-dependent expression of pro-IL-1β and NLRP3. In the second stage, a different signal, which may include cytosolic K^+ reduction, reactive oxygen species, or lysosomal destabilization, triggers the binding of NLRP3 to apoptosis-associated speck-like protein containing a CARD domain (ASC) and sequential recruitment of procaspase-1. The successful formation of NLRP3-ASC-procaspase-1 complexes allow the processing of procaspase-1 into

Francesco Di Virgilio and Pablo Pelegrín (eds.), *NLR Proteins: Methods and Protocols*, Methods in Molecular Biology, vol. 1417, DOI 10.1007/978-1-4939-3566-6_12, © Springer Science+Business Media New York 2016

active caspase-1 [1–6], which is essential for the cleavage of pro-IL-1β and pro-IL-18 to produce mature IL-1β and IL-18. NLRP3 inflammasome activation is associated with many chronic diseases including familial cold autoinflammatory syndrome, gout, silicosis, asbestosis, atherosclerosis, and metabolic diseases [2–6]. Despite the critical pathological role of the inflammasome, the exact molecular processes involved in the activation of NLRP3 inflammasome remain incompletely understood.

A few methods have been developed to measure the oligomerization of NLR (*see* Chapters 8–11 in this book). Due to the atypical nature of NLRP3 inflammasome assembly, its detection is more complicated than other protein complexes. One difficulty in biochemically monitoring the formation of the NLRP3 inflammasome arises from the fact that cell lysis, considered a danger signal due to cell membrane damage, induces spontaneous inflammasome assembly [1]. Incubation of cell lysates from TLR-primed macrophages at 30 °C leads to spontaneous NLRP3 inflammasome formation within 15 min in the absence of the second signals [1, 7] (*see* Chapter 14 in this book). This limits monitoring of the NLRP3 inflammasome assembly by immunoprecipitation of macrophage cell lysates. Another unique feature of NLRP3 inflammasome is its structure: NLRP3 nucleates ASC oligomerization and ASC filaments and in turn serves as a template for procaspase-1 polymerization to form a star-shaped complex, in which ASC constitutes the center and procaspase-1 forms the arms [8]. The stoichiometry of each component in NLRP3-ASC-procaspase-1 complex is highly disproportional, and very few NLRP3 are present in the ternary complex while ASC is under-stoichiometric to procaspase-1. The unusual stoichiometry of the NLRP3 inflammasome may explain the difficulties demonstrating the co-localization of NLRP3 with ASC or caspase-1 in inflammasome formation.

The proximity ligation assay (PLA) measures protein–protein interaction in a cellular environment based on the formation of a reporter molecule facilitated by the proximal binding of two DNA probes. This approach is different in principle from the commonly used methods of co-immunoprecipitation, co-localization under microscopy, or fluorescence resonance energy transfer, and has been used as an alternative method of studying protein-protein interactions [9, 10]. Similar to many other methods of studying complex formation, the proteins of interest (using NLRP3 and ASC as an example) are first bound by specific primary antibodies (Fig. 1a). A requirement is that the two primary antibodies come from two different species, selecting from mouse, rabbit, or goat. This enables the binding of the two secondary antibodies (e.g., anti-mouse and anti-rabbit) that have been conjugated with different DNA probes (Fig. 1b). If NLRP3 and ASC are in close proximity (<40 nm), the two DNA probes will situate close together, allowing hybridization with two additional oligonucleotides

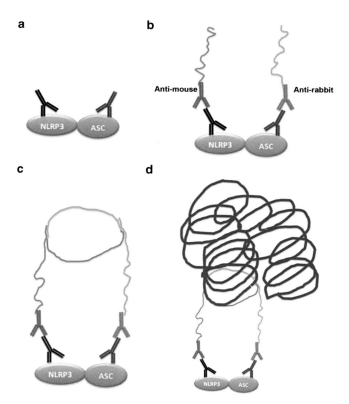

Fig. 1 Measurement of NLRP3-ASC association by in situ PLA. (**a**) Binding of primary antibodies. NLRP3 and ASC are first bound by mouse anti-NLRP3 and rabbit anti-ASC. (**b**) Binding of secondary antibodies. This is followed by the binding of anti-mouse and anti-rabbit secondary antibodies, each attached to a different DNA probe (PLUS probe and MINUS probe, blue and green in the figure). (**c**) Hybridization and ligation. The close promixity between NLRP3 and ASC allows the hybridization of the two DNA probes with two additional DNA oligonucleotides (*orange* and *red*) to form a DNA bridge between NLRP3 and ASC after ligation. (**d**) Amplification. The DNA link serves as the template for rolling-circle amplification by Ø29 DNA polymerase, and the amplified DNA product is visualized after hybridizing with fluorescence labeled probe

containing regions complementary to each of the two antibody-linked DNA probes (Fig. 1c). The ligation of the two oligonucleotides to antibody-attached DNA probes form a circular DNA bridge between the two secondary antibodies. The circular DNA then serves as a template for rolling-circle amplification by polymerase (Fig. 1d). The amplified DNA products are detected by hybridizing with fluorescent-labeled oligonucleotides and are visualized as a spot by fluorescence microscopy (Fig. 2). In this protocol, we describe the measurement of NLRP3 inflammasome assembly by in situ PLA in macrophages activated through a combination of LPS and ATP. Since NLRP3 inflammasome assembly is stimulated by inflammatory, infectious, or danger signals, the

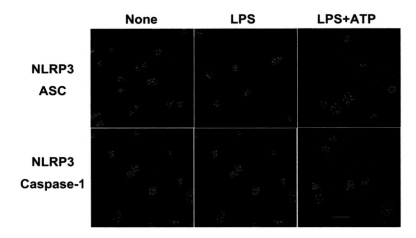

Fig. 2 Detection of NLRP3-ASC and NLRP3-procaspase-1 in situ interaction in macrophages. Bone marrow-derived macrophages were left untreated or treated with LPS or LPS plus ATP. Untreated and LPS alone were used as negative controls. Cells were washed, fixed, and permeabilized. Macrophages were then incubated with primary antibodies (mouse anti-NLRP3, rabbit anti-ASC, or rabbit anti-caspase-1) overnight at 4 °C. Cells were labeled with PLUS probe-conjugated anti-mouse antibodies and MINUS probe-conjugated anti-rabbit antibodies. After annealing and ligation with the two template oligonucleotides, the circularized template was amplified. The amplified sequence was detected by hybridization with a Texas Red-labeled probe. The cells were examined under a Zeiss LSM 780 confocal laser scanning microscope with a 63× objective lens. Texas Red-linked oligonucleotide was detected using a 598 nm (Ex) and 613 nm (Em) filter set, while DAPI was detected using a 360 nm (Ex) and 460 nm (Em) filter set. Bar indicates 20 μm

current protocol could be extended to study inflammasome formation induced by these stimuli. In situ PLA allows us to detect not only NLRP3 inflammasome complex formation in an intact cell or tissue sample [7, 11–13], but also the involvement of other proteins in the processes of NLRP3 complex formation [7, 11].

2   Materials

2.1   *Cell Culture*

1. Bone marrow-derived macrophages (BMDMs): Bone marrow cells collected from tibias and femurs by flushing with cold PBS with a 25G needle cultured for 6 days in complete DMEM medium: DMEM with 10 % fetal bovine serum, 10 mM glutamine, 100 U/ml penicillin, 100 μg/ml streptomycin, and 20 μM 2-mercaptoethanol (2-ME), with 20 % L929 conditioned media to generate BMDMs (*see* **Note 1**).

2. THP-1 cells (American Type Culture Collection): Human monocyte cell line THP-1 (*see* **Note 2**) cultured in complete RPMI medium: RPMI1640 supplemented with 10 % fetal bovine serum, 10 mM glutamine, 100 U/ml penicillin, 100 μg/ml streptomycin, and 20 μM 2-ME.

3. 12-O-tetradecanoylphorbol-13-acetate (TPA, also known as PMA).

4. LPS ultrapure from *E. coli O111:B4.*

5. Adenosine 5′-triphosphate disodium salt hydrate (ATP).

6. Nigericin.

7. 12-well plates.

8. 3.5 mm coverslips sterilized (autoclaved or by dipping in 70 % ethanol).

2.2 Fixation and Permeabilization

1. Phosphate buffered saline (PBS): 137 mM NaCl, 2.7 mM KCl, 1 mM Na_2HPO_4, 2 mM KH_2PO_4, pH 7.4.

2. Fixation solution: 4 % formaldehyde in PBS. Fixation solution should be freshly prepared each time.

3. Permeabilization solution: 0.1 % Triton X-100 in PBS.

4. Mounting medium.

5. Super PAP Pen Liquid Blocker (TED PELLA, Redding, CA, USA).

2.3 Primary Antibodies

1. Mouse anti-mouse NLRP3 antibody (Cryo2, ENZO Life Science).

2. Mouse anti-human NLRP3 antibody (nalpy3-b, ENZO Life Science).

3. Rabbit anti-ASC antibody (AL177, ENZO Life Science).

4. Rabbit anti-mouse caspase-1 antibody (M-20, Santa Cruz Biotech).

5. Rabbit anti-human caspase-1 (A-19, Santa Cruz).

2.4 Duolink In Situ Reagents

All reagents are exclusive from Olink Bioscience and distributed by Sigma-Aldrich.

1. PLA probes: 1× Blocking Solution, 1× Antibody Diluent, 5× PLA probe anti-mouse MINUS, and 5× PLA probe anti-rabbit PLUS.

2. Detection reagents: 5× ligation stock, 1 U/μl ligase, 5× Amplification Red, and 10 U/μl polymerase.

3. Mounting medium with DAPI.

4. The above reagents are available together in a starter kit (*see* **Note 3**) or can be purchased separately.

2.5 Wash

1. Staining jar.

2. Wash buffer A: 150 mM NaCl, 10 mM Tris–HCl, 0.05 % Tween 20, pH 7.4. Filter the solution through a 0.22 μm filter and store at 4 °C. Warm to room temperature before use.

3. Wash buffer B: 100 M NaCl and 20 M Tris–HCl, pH 7.4. Filter the solution through a 0.22 μm filter and store at 4 °C. Warm to room temperature before use.

3 Methods

3.1 Culture and Inflammasome Activation of Bone Marrow-Derived Macrophages

1. In tissue culture hood, place sterile 3.5 mm coverslips into 12-well plate.

2. Add 2×10^5 bone marrow-derived macrophages (BMDMs) in 0.5 ml complete DMEM medium to each well and leave cells to adhere to coverslip for 2 h (*see* **Note 4**). The attachment, priming, and activation of macrophages (**steps 2–4**) are performed at 37 °C in a CO_2 incubator.

3. Prime BMDMs with 0.5 ml 1 μg/ml LPS for 3 h (*see* **Note 5**).

4. Replace LPS with 0.5 ml 2.5 mM ATP without washing. Activate BMDMs with ATP for 15 min (*see* **Notes 6–8**).

5. Stop the activation by washing cells immediately with 0.5 ml of 1× PBS twice (*see* **Note 9**).

3.2 Culture and Inflammasome Activation of THP-1 Cells

1. In tissue culture hood, place sterile 3.5 mm coverslips into 12-well plate.

2. Add 5×10^5 THP1 cells in 0.5 ml complete RPMI medium containing 500 ng/ml TPA to each well. After 3 h TPA priming, remove TPA-containing medium and replace with fresh RPMI medium. Incubate overnight for differentiation into macrophages (*see* **Note 10**). The differentiation and activation of THP-1 cells (**steps 2 and 3**) are performed at 37 °C in a CO_2 incubator.

3. Activate THP1 cells with 0.5 ml of 1 mM ATP for 30 min (*see* **Notes 6** and **7**).

4. Stop the activation by immediately and quickly washing cells twice with 0.5 ml of 1× PBS (*see* **Note 9**).

3.3 Cell Fixation

1. Fix cells by adding 0.5 ml of freshly prepared fixation solution to each well (*see* **Note 11**) and incubate for 30 min at room temperature with gentle shaking on a lab shaker.

2. Wash cells with 0.5 ml PBS per well for 5 min with gentle shaking. Repeat wash two more times.

3. Permeabilize cells with 0.5 ml of permeabilization solution for 3 min at room temperature with gentle shaking.

4. Wash cells with 0.5 ml of PBS for 5 min with gentle shaking. Repeat wash additional two more times.

5. Wash cells with distilled water once to remove salt. Remove slides from 12-well plate and place on paper towel. Allow slides to dry at room temperature for 20 min.

6. With sample side up, adhere the coverslip to the slide firmly with Mounting Medium (*see* **Note 12**).

7. Mark and seal the reaction area using a Super PAP Pen Liquid Blocker (*see* **Note 13**).

8. Check cell morphology under a microscope, and proceed with proximity ligation assay only if cells are intact and healthy-looking (*see* **Note 14**).

3.4 Proximity Ligation Assay

1. Add 35 μl of Blocking Solution to cells and incubate in humidity chamber for 30 min at room temperature (*see* **Note 15**).

2. During the incubation, prepare the primary antibody dilution. For determining the association between NLRP3 and caspase-1 in mouse macrophages, dilute antibodies in 1× Antibody Diluent to generate mixture containing 10 μg/ml of anti-mouse NLRP3 (Cryo2) and 0.2 μg/ml of anti-mouse caspase-1 (M-20) (*see* **Notes 16** and **17**).

3. For measuring the association between NLRP3 and caspase-1 in THP1 cells and human macrophages, dilute antibodies in 1× Antibody Diluent to generate mixture containing 10 μg/ml of anti-human NLRP3 (nalpy3-b) and 0.2 μg/ml of anti-human caspase-1 (A-19).

4. For detecting the association between NLRP3 and ASC, dilute both anti-NLRP3 (Cryo2 or nalpy3-b) and anti-ASC (AL177) to final 10 μg/ml each.

5. Tap off the Blocking Solution from the slides (*see* **Note 18**).

6. Add 35 μl of diluted primary antibody solution to each cell sample and incubate slides overnight in humidity chamber at 4 °C (*see* **Note 19**).

7. Dilute the 5× PLA probe anti-mouse MINUS and 5× PLA probe anti-rabbit PLUS in Antibody Diluent to generate mixtures containing 1× of each probe (*see* **Note 20**).

8. Tap off the primary antibody solution from the slides.

9. Wash slides in staining jar twice with PBS for 5 min with gentle orbital shaking (*see* **Notes 21** and **22**).

10. Add 35 μl of diluted probe solution to each cell sample and incubate for 1 h in humidity chamber preheated to 37 °C.

11. Dilute 5× ligation stock with molecular biology-grade pure water (e.g., autoclaved MilliQ water) to make 1× ligation stock (*see* **Note 23**), leaving the exact volume sufficient for the addition of ligase to make 1:40 dilution in **step 14**.

12. Tap off the probe solution from the slides.

13. Wash slides in staining jar with wash buffer A for 5 min × 2 with gentle shaking.

14. Remove ligase from –20 °C freezer and dilute 1:40 with 1× ligation stock (**step 10**). Add 35 µl of diluted ligase solution to each sample and incubate for 30 min in humidity chamber at 37 °C.

15. Dilute 5× amplification stock with molecular biology grade pure water to make 1× amplification stock, with the exact volume sufficient for the addition of polymerase to make 1:80 dilution in **step 18**.

16. Tap off the ligation solution from the slides.

17. Wash slides in staining jar with 1× wash buffer A for 2 min × 2 with gentle shaking.

18. Remove polymerase from –20 °C freezer and dilute 1:80 in 1× Amplification Red (*see* **Note 24**). Add 35 µl of diluted amplification solution to each sample and incubate for 100 min in humidity chamber at 37 °C.

19. Tap off the amplification solution from the slides.

20. Wash slides twice in staining jar with 1× wash buffer B for 10 min. Next wash slides in staining jar with wash buffer B diluted 1:100 (final 0.01×) for 1 min.

21. Tap off the excess wash buffer. Allow slides to air-dry at room temperature for 5 min in the dark (*see* **Note 25**).

22. Add 6 µl of the mounting medium with DAPI and then apply a coverslip on top of each sample (*see* **Note 26**).

23. Analyze samples with a fluorescence or confocal microscope (Fig. 2).

24. Store the slides at –20 °C in the dark.

25. Quantitate the number of the spots in each cell (*see* **Note 27**).

4 Notes

1. NLRP3 inflammasome activation can also be studied in bone marrow-derived dendritic cells and peripheral blood-derived human macrophages following the protocol for mouse bone marrow-derived macrophages.

2. Relative to peripheral blood-derived macrophages, THP-1 cells possess advantageous traits such as a homogenous genetic background, higher growing rates, and availability in large quantities [14]. Heightened transfection efficiency allows specific genes to be more effectively knocked down or overexpressed in THP-1 cells than normal macrophages in

inflammasome study. THP-1 with ASC knockdown or NLRP3 knockdown are also available from InvivoGen. However, it is important to note that the malignant nature of THP-1 cells may generate results atypical from normal macrophages.

3. Available in Duolink In Situ Red Starter kit or Duolink In Situ Orange Starter kit. Each kit contains PLA probe anti-mouse MINUS, PLA probe anti-rabbit PLUS, wash buffers for fluorescence, and mounting medium with DAPI, with choice of Red (Texas Red) or Orange (Cy3) Detection Reagents.

4. Optimal cell confluence is ~80 % at the time of experiment. Fully confluent macrophages are not suitable for in situ PLA examination due to the potential variability in NLRP3 inflammasome activation and the generation of overlapping image.

5. LPS priming represents the first signal required for the upregulation of pro-IL-1β and NLRP3 in BMDMs. LPS could be substituted by other TLR agonists such as Pam3CSK4 (for TLR4) or R848 (for TLR7/8).

6. ATP is the second signal triggering the formation of NLRP3-ASC-procaspase-1 inflammasome. Nigericin (10 μM) is an alternative second signal activator for NLRP3 inflammasome assembly. Monosodium urate (MSU) and alum crystals are other compounds commonly used to activate NLRP3 inflammasome [3, 4]; however, due to the crystalline nature, they are more difficult to wash off from macrophages after activation and may interfere with microscopic observation.

7. The concentration of ATP and the time of incubation for optimal inflammasome detection should be tested. Inflammasome activation is followed by pyroptosis [15], and incubation with ATP for too long will lead to diminished cell viability (*see* Chapter 17 in this book).

8. Untreated macrophages and macrophages stimulated with LPS only served as the negative controls for NLRP3 activation by in situ PLA (Fig. 2).

9. Limit time in the wash step to minimize cell death.

10. For THP1 cells, TPA treatment will increase the expression of pro-IL-1β and NLRP3, and LPS priming is not essential.

11. Freshly dilute paraformaldehyde to final concentration of exactly 4 % in 1× PBS. The concentrations of PBS and paraformaldehyde are critical for intact cell fixation.

12. When adhering the coverslip to slide, keep the coverslip horizontal to slide. This ensures the sample will remain in focus in an optic field under microscopic observation. Make sure the mounting medium is dry before proceeding.

13. Liquid blocker must be uniformly applied around the sample area to prevent reagent leakage and sample over drying.

14. Do not proceed with the proximity ligation assay if fixed cells do not appear morphologically robust, since the PLA reagents are relatively expensive and success is highly dependent on the quality of the fixed macrophages.

15. The blocking solution should cover the entire area with cells.

16. The two primary antibodies should come from different species out of mouse, rabbit, and goat, and their isotypes must be IgG. Affinity purified polyclonal antibodies are preferred over unpurified polyclonal antibodies.

17. The choice of the primary antibodies is critical to PLA. For use of antibodies to detect the association of the other molecules with NLR, first validate the recognition ability and titer by immunofluorescence staining or intracellular staining.

18. Never allow samples to dry out before the addition of the primary antibodies.

19. Make sure the humidity chamber works. Do not allow samples to dry out.

20. For in situ PLA to detect the association of other proteins, the use of secondary antibodies will be based on the species of primary antibodies, but the secondary antibodies must be in the pair combination of PLA probe MINUS and PLA probe PLUS.

21. Each wash in a small staining jar requires about 150 ml PBS.

22. PBS is used for washing off the primary antibodies in this protocol. Selection of wash buffer for other primary antibodies should be based on the antibodies used.

23. The ligation stock contains two additional oligonucleotides to hybridize with the two DNA probes attached to secondary antibodies (Fig. 1c).

24. The amplification stock contains the all constituents required for rolling circle amplification and for the fluorescence-labeled oligonucleotides to hybridize with the amplified DNA product. Four different fluorescence detection reagents are available from Olink Bioscience: Green (FITC-like), Orange (Cy3-like), Red (Texas Red-like), and Far Red (Cy5-like).

25. Do not allow the slides to completely dry out. One can alternatively use Kimwipes tissue paper to absorb the excess liquid from slides. Proceed with next step immediately.

26. Slowly apply the coverslip on top of samples to avoid bubbles between two coverslips.

27. Image analysis software (Duolink ImageTool) for quantitation of PLA products can be purchased from Olink Bioscience. Unless the sample size is large, counting PLA signal for 100–200 cells per treatment can also be managed manually.

Acknowledgements

We thank Dr. AndreAna Peña for manuscript editing. This work was supported by grant MOST103-2321-B001-031 from the National Science Council, and an Academia Sinica Investigator Award from Academia Sinica, Taiwan, R.O.C.

Conflict of interest: The authors declare no conflict of interest.

References

1. Martinon F, Burns K, Tschopp J (2002) The inflammasome: a molecular platform triggering activation of inflammatory caspases and processing of proIL-1β. Mol Cell 10:417–426

2. Mariathasan S, Weiss DS, Newton K et al (2006) Cryopyrin activates the inflammasome in response to toxins and ATP. Nature 440:228–232

3. Martinon F, Petrilli V, Mayor A et al (2006) Gout-associated uric acid crystals activate the NLRP3 inflammasome. Nature 440:237–241

4. Rock KL, Latz E, Ontiveros F, Kono H (2010) The sterile inflammatory response. Annu Rev Immunol 28:321–342

5. Lamkanfi M, Dixit VM (2014) Mechanisms and functions of inflammasomes. Cell 157:1013–1022

6. Robbins GR, Wen H, Ting JP (2014) Inflammasomes and metabolic disorders: old genes in modern diseases. Mol Cell 54:297–308

7. Wu YH, Kuo WC, Wu YJ et al (2014) Participation of c-FLIP in NLRP3 and AIM2 inflammasome activation. Cell Death Differ 21:451–461

8. Lu A, Magupalli VG, Ruan J et al (2014) Unified polymerization mechanism for the assembly of ASC-dependent inflammasomes. Cell 156:1193–1206

9. Söderberg O, Gullberg M, Jarvius M et al (2006) Direct observation of individual endogenous protein complexes in situ by proximity ligation. Nat Methods 3:995–1000

10. Koos B, Andersson L, Clausson CM et al (2014) Analysis of protein interactions in situ by proximity ligation assays. Curr Top Microbiol Immunol 377:111–126

11. Chuang YT, Lin YC, Lin KH et al (2011) Tumor suppressor death-associated protein kinase is required for full IL-1β production. Blood 117:960–970

12. Lo YH, Huang YW, Wu YH et al (2013) Selective inhibition of the NLRP3 inflammasome by targeting to promyelocytic leukemia protein in mouse and human. Blood 121:3185–3194

13. Misawa T, Takahama M, Kozaki T et al (2013) Microtubule-driven spatial arrangement of mitochondria promotes activation of the NLRP3 inflammasome. Nat Immunol 14:454–460

14. Chanput W, Mes JJ, Wichers HJ (2014) THP-1 cell line: an in vitro cell model for immune modulation approach. Int Immunopharmacol 23:37–45

15. Fernandes-Alnemri T, Wu J, Wu JW et al (2007) The pyroptosome: a supramolecular assembly of ASC dimers mediating inflammatory cell death via caspase-1 activation. Cell Death Differ 14:1590–1604

Chapter 13

Assessing Caspase-1 Activation

Baptiste Guey and Virginie Petrilli

Abstract

The caspase-1 enzymatic activity plays a major role in the innate immune response as it regulates the maturation of two major proinflammatory cytokines, the interleukin-1beta (IL-1β) and IL-18. In this chapter, we describe the technique of Western blot to assess caspase-1 activation. This method provides multiple information within one experiment. It allows the detection of both unprocessed and processed caspase-1 and substrates.

Key words Western blot, Inflammasome, Interleukin-1beta (IL-1β), Caspase-1

1   Introduction

Caspase-1 is an enzyme belonging to a family of cysteine proteases that cleave their substrates after an aspartic acid residue. Like apoptotic initiator caspases (e.g., caspase-8 or caspase-9), it is synthesized as a zymogen. Caspase-1 consists of three main domains, a Caspase Activation and Recruitment Domain (CARD) at the N-terminus of the protein followed by two catalytic subunits, the p20 and the p10 (Fig. 1). As an initiator caspase, caspase-1 gets activated within a multi-protein complex, named the "inflammasome." The inflammasome acts as a molecular platform inducing the zymogen dimerization to initiate caspase-1 activation by autoproteolysis. The freed subunits assemble into dimers of p20/p10 to form the active caspase-1 [1, 2]. The best-characterized substrates of caspase-1 are the proinflammatory cytokines, the pro-interleukin-1β (pro-IL-1β) and pro-IL-18. Their cleavage by caspase-1 results in the production and secretion of the biologically active cytokines IL-1β and IL-18. Of note, upon inflammasome activation, the p20/p10 dimers of caspase-1 are secreted together with the mature substrates. This is the reason why cell supernatants are analyzed by Western blot.

Inflammasomes are major actors of the innate immune response. They are expressed by myeloid cells such as macrophages

Francesco Di Virgilio and Pablo Pelegrín (eds.), *NLR Proteins: Methods and Protocols*, Methods in Molecular Biology, vol. 1417, DOI 10.1007/978-1-4939-3566-6_13, © Springer Science+Business Media New York 2016

Caspase-1 zymogen

Active dimers of Caspase-1

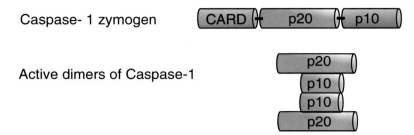

Fig. 1 Schematic representation of the full length caspase-1, size 45 kDa, and of the active dimers

and dendritic cells, but also by epithelial cells such as keratinocytes or intestinal cells. Different cytosolic pattern recognition receptors such as nucleotide-binding domain and leucine rich repeat pyrin containing 1 (NLRP1), NLRP3, NLRC4, and absent in melanoma 2 (AIM2) are able to trigger the assembly of an inflammasome upon sensing of specific pathogen-associated molecular pattern (PAMP) or danger-associated molecular pattern (DAMP). The NLRP3 receptor is able to detect a wide range of DAMP, e.g., extracellular ATP and monosodium urate crystals, and PAMP, e.g., *Staphylococcus aureus*, influenza virus, adenovirus and bacterial toxins such as nigericin. The mouse NLRP1b is activated by the lethal toxin of *Bacillus anthracis*, while the NLRC4 detects bacteria like *Salmonella* or *Shigella* [3]. Inflammasome activation is often associated with a specific form of cell death dependent on caspase-1 activity called pyroptosis.

A convenient method to monitor caspase-1 activation is the detection of the autoproteolysis of the protein using the technique of Western blotting. Separation of the different forms of caspase-1 protein using denaturing polyacrylamide gel electrophoresis followed by Western blot allows the visualization of the full length caspase-1 at 45 kDa (zymogen) and of the cleaved fragments of the protein resulting from its autoprocessing, for instance p20 or p10 depending on the epitope recognized by the antibody. Similarly this technique allows the detection of both the immature and mature form of the substrate of caspase-1, for instance the pro-IL-1β at 35 kDa and the mature IL-1β at 17 kDa. However, recent studies identified specific situations where caspase-1 is active despite the absence of autoprocessing. Thus, for some situations, to assess caspase-1 activity, monitoring its autoprocessing is not sufficient and assessing substrate cleavage and the induction of cell death are also required [4, 5]. We therefore include a simple method to monitor cell death, the measurement of lactate dehydrogenase released in the medium.

In this chapter, we activate caspase-1 by triggering the formation of the NLRP3 and NLRP1b inflammasomes in mouse peritoneal macrophages. This method can be applied to bone marrow derived macrophages and dendritic cells, and even to human macrophages, using adequate antibodies. It can also be applied to monitor the activation of other inflammasomes using cognate activators.

2 Materials

2.1 Cells and Inflammasome Stimulation

All buffers and solutions must be prepared using ultrapure water.

1. Mouse peritoneal macrophages from WT, and ideally, from *caspase-1* deficient (KO) mice that will serve as negative control (for method *see* ref. 5). If no caspase-1 KO mice are available, the caspase-1 inhibitor Z-YVAD-fmk can be used.

2. 12-well plates treated for tissue culture.

3. Dulbecco's modified Eagle's medium (DMEM) cell culture medium complemented with 10 % (vol/vol) decomplemented fetal bovine serum, 1 mM sodium pyruvate, 100 IU/ml penicillin/streptomycin, and 2 mM glutamine.

4. Opti-MEM® medium (this medium is exclusive of Life Technologies); other serum free media could be also used.

5. Phosphate buffered saline solution (PBS): 137 mM NaCl, 2.7 mM KCl, 1 mM Na_2HPO_4, 2 mM KH_2PO_4, pH 7.4.

6. Ultrapure *Escherichia coli* 0111:B4 lipopolysaccharide (LPS).

7. Stock solution of 10 mM nigericin in ethanol.

8. 1 mM adenosine triphosphate (ATP) stock solution in water (*see* **Note 1**).

9. Lethal toxin (LT) from anthrax: the stock is composed of 1 µg/µl of lethal factor mixed with 1 µg/µl of protective antigen [5].

10. 6 mM EDTA in PBS buffer: dilute 0.5 M EDTA stock solution in PBS.

2.2 Protein Extractions

1. RIPA lysis buffer stock 2×: Mix 5 ml of 1 M Tris–HCl pH 7.5, 3 ml 5 M of NaCl, 10 ml of 10 % sodium deoxycholate, 0.15 ml of 20 % SDS, 1 m of 0.5 % Triton X-100, 1 ml of 0.5 M EDTA, and 0.2 ml of 1 M sodium orthovanadate. Add two tablets of Roche protein inhibitor cocktail; other protein inhibitors cocktails are also suitable. Make the volume up to 50 ml with H_2O. Mix well. Aliquot 0.5 ml in 1.5 ml microtubes and store at –80 °C.

2.3 SDS-PAGE

1. Bio-Rad mini-PROTEAN Tetra cell electrophoretic system (or equivalent).

2. Glass plates with 1.5 mm spacer, 15-well combs.

3. Solution of premixed 30 % ratio 37.5:1 acrylamide–bis-acrylamide.

4. 10 % (wt/vol) ammonium persulfate (APS) solution in water.

5. N,N,N,N'-tetramethyl-ethylenediamine (TEMED).

6. Stacking gel buffer: 0.5 M Tris–HCl pH 6.8: Weigh 60 g of tris base, add 0.5 L of H_2O, mix using magnetic stirrer, adjust pH to 6.8 using HCl (*see* **Note 2**) and add H_2O up to 1 L. Store at +4 °C.

7. Separating gel buffer 1.5 M Tris–HCl pH 8.8: Weigh 181.5 g tris base, add 0.7 L of H_2O, mix, adjust pH to 8.8 using HCl (*see* **Note 2**), make up to 1 L with H_2O. Store at +4 °C.

8. Separating gels: for two 15 % acrylamide mini-gels mix 4.8 ml of H_2O, 10 ml of 30 % acrylamide–bis-acrylamide, 5 ml of separating gel buffer, 0.1 ml of 20 % SDS, 0.2 ml of 10 % APS, 0.01 ml TEMED.

9. Stacking gels: for two 5 % acrylamide mini-gels mix 5 ml H_2O, 1.7 ml of 30 % acrylamide–bis-acrylamide, 2.5 ml of staking gel buffer, 0.05 ml of 20 % SDS, 0.1 ml of 10 % APS, 0.01 ml TEMED.

10. Running (migration) buffer 10× stock solution: Weigh 60.6 g of tris base, 288 g of glycine, add 0.1 L of 20 % SDS, make up to 2 L with H_2O. Mix.

11. 3× Laemmli buffer (LB): Weigh 60 mg phenol red (*see* **Note 3**), add 69 ml of glycerol, 37.5 ml of 1 M Tris–HCl pH 6.8 and 60 ml of 20 % SDS. Make up to 200 ml using H_2O. Mix well. Verify final pH is 6.8 and adjust if necessary. Just before adding the LB to the protein extracts, add 100 mM DTT to reach final concentration.

12. 1 M dithiothreitol (DTT): Dissolve 15.45 g of DTT in H_2O, aliquot into microtubes and store at –20 °C.

13. Protein ladder.

2.4 Western Blot and Antibody Incubation

1. Bio-Rad Mini Transblot electrophoretic system (or equivalent).

2. Transfer buffer × 1 (store at +4 °C): weigh 25 g of tris base, 120 g of glycine, add 8 L of H_2O, mix. Adjust pH to 8.3 using HCl, add 1.6 L of pure ethanol, make up the volume to 10 L with H_2O.

3. 10× stock solution of tris buffer saline Tween (TBST) (store at +4 °C): 500 ml of 1 M tris base pH 7.5, 10 ml of Tween 20, 300 ml of 5 M NaCl, and 190 ml of H_2O.

4. Nitrocellulose membranes with 0.22 μm pore size.

5. Whatman paper, cut into 7.5 × 7.5 cm pieces.

6. Ponceau S.

7. Nonfat dry milk.

8. Blocking buffer: 5 % (wt/vol) nonfat dry milk in TBST 1×.

9. Primary antibodies: 1:500 dilution of anti-caspase-1 Casper-1 mouse (Adipogen), 1:1000 dilution of anti-IL-1β IL-1F2 (R&D Systems).

10. Secondary antibodies coupled to the horseradish peroxidase (HRP): 1:5000 dilution of donkey anti-goat or donkey anti-mouse.

11. Chemoluminescence substrate for HRP.

12. Frozen ice pack for the transfer.

2.5 Cell Death Assay

1. Lactate dehydrogenase (LDH) activity assay kit, such as the CytoTox 96® Non-Radioactive Cytotoxicity Assay (Promega). However, other brands are also suitable. Prepare LDH substrate mix following manufacturer's instructions.

2. Flat-bottom 96-well plate.

3 Methods

3.1 Activation of the Inflammasome

1. Plate the WT and *caspase-1* KO macrophages at a density ranging from 8×10^5 to 10^6 cells in a 12 well-plate in 1 ml of complete DMEM. Make sure to seed the adequate number of wells; do not forget to seed an extra control well that will be used as a positive control for the LDH assay. After 2 h, change the medium to remove the non-adherent cells.

2. Prime the cells overnight by stimulating them in complete DMEM containing 0.5 μg/ml of LPS. This step allows the induction of the production of the pro-IL-1β and of the NLRP3 proteins.

3. The following morning, remove the medium and replace it with 1 ml of Opti-MEM (*see* **Note 4**). Then stimulate the cells to activate the inflammasome by adding either 10 μM nigericin per well for 2–4 h or 5 mM ATP for 30 min or 1 μg/ml LT for 6 h, or the solvent to the control wells. If using Z-YVAD-fmk, add the inhibitor at 50 μM 30 min prior to the stimulation.

3.2 Sample Preparation

1. Collect each 1 ml of cell supernatant into labeled microtubes and centrifuge gently at $1500 \times g$ to get rid of any dead cells. Remove the supernatant and store it in 2 aliquots of 0.5 ml at −20 °C (1 for protein precipitation and 1 for other applications such as LDH assay or ELISA)

2. Wash the cells with PBS and then incubate the cells in 0.5 ml of 6 mM EDTA in PBS for 5 min on ice. Scrape the cells and collect the cells into a clean microtube. Pellet the cells at $1500 \times g$ in a refrigerated centrifuge. Discard the supernatant, store the dry cell pellet at −80 °C or proceed directly to the cell lysis.

3. Cell extracts: Make the RIPA buffer 1× by adding 0.5 ml of water, and lyse the cells in 40 μl of buffer. Incubate on ice for 30 min and centrifuge at +4 °C at 15,800×g. Collect the cell protein extract into a clean tube (optional: measure the protein concentration using the Bradford assay and use bovine serum albumin as a standard). Add Laemmli buffer containing DTT to the extracts, heat for 5 min at 95 °C. Load directly onto SDS-PAGE or store at −80 °C.

4. Supernatants: to analyze the protein contents of cell supernatants, we use the methanol/chloroform protein precipitation method. This part should be conducted under a fume hood. To 0.5 ml of cell supernatant, add 0.1 ml of chloroform and 0.5 ml of methanol. Vortex. Centrifuge for 3 min at 15,800×g. The white protein disk is visible at the interphase between the aqueous and organic phase. Discard most of the upper phase without disturbing the protein pellet. Add 0.5 ml of methanol, vortex, and centrifuge for 3 min at 15,800×g. The protein pellet should be stacked at the bottom of the tubes. Discard the supernatant. Leave the pellets to air-dry under a fume hood to remove any trace of solvent. Resuspend the proteins in 50 μl of LB 3× containing DTT (*see* **Note 5**). Heat at 95 °C for 5 min. Load directly onto gels or store at −80 °C.

3.3 SDS-Polyacrylamide Gel Casting and Electrophoresis

1. Assemble the glass plates (1.5 mm spacer) using the casting system of miniPROTEAN electrophoretic system.

2. Prepare Separating gels.

3. Mix well.

4. Pour 7.5–8 ml of gel preparation into the plates and gently overlay the gel with 0.5 ml water. Leave to polymerize (20 min).

5. Prepare the stacking gels.

6. Mix well.

7. Remove the water from the surface of the gels and pour the stacking gels on top of the separating gels. Quickly add the combs to the stacking gel. Wait until gels are polymerized (*see* **Note 6**).

8. Install the glass plates containing the gels into the migration tank, fill with running buffer 1× and gently remove the combs.

9. Load the gels: add the protein ladder to one well and load 20 μg of samples (or 15–20 μl) per well. Top up the tank with running buffer 1× if necessary (*see* **Note 7**).

10. Start the electrophoresis at 70 V for 20 min then increase the voltage to 100 V for 90 min. Regularly check the migration. Another method is to electrophorese by applying 35 mA/gel.

3.4 Western Blot, Wet Method

1. Prepare a tank containing transfer buffer and soak the foam pads and membranes in this buffer.

2. Uncast the gels and remove the gels from the glass plates carefully.

3. Prepare the transfer sandwich by stacking 1 foam pad, 2 Whatman papers soaked into the transfer buffer, the gel (we remove the stacking gel before the transfer), the nitrocellulose membrane, and 2 Whatman papers soaked in the transfer buffer, on the black lid of the transfer cassette (Fig. 2). Get rid of any bubbles by gently rolling the top of the pile with half of a 5 ml pipette. Add another foam pad and close the cassette, place it into the transfer cell, taking care to put the cassette in the proper orientation with regards to the electrodes (with the Bio-Rad system the black side of the cassette toward the black side of the transfer cell). Add one ice block to the transfer tank and fill with transfer buffer (*see* **Note 8**).

4. Transfer at 100 V for 1 h.

3.5 Probing and Detection

1. Open the transfer cassettes to collect the membranes.

2. Optional: you may want to incubate your membrane in a Ponceau S solution before the blocking step. The Ponceau S stains the proteins and is useful to visualize the quality of the protein transfer and the protein load of each well. Incubate the membranes in a solution of Ponceau S for 5 min under agitation, then rinse briefly several times with water.

3. Block the membranes by incubating them in the blocking buffer for 1 h at room temperature and under gentle agitation.

4. Incubate the membranes overnight in the primary antibodies diluted in the blocking buffer at +4 °C under gentle agitation.

5. Proceed to the washing step: add at least 10 ml of TBST 1× for 15 min at room temperature under agitation. Discard the buffer and repeat this step three times.

6. Incubate the membranes in the secondary antibody solution. Dilute the appropriate horseradish-coupled antibody in 1 % milk TBST.

7. Wash the membranes with 10 ml of TBST 1× for 15 min at room temperature under agitation. Discard the buffer and repeat this step three times.

8. Incubate the membranes in HRP substrate. For detection of IL-1β and caspase-1, we incubate the membranes for 5 min in the luminescence HRP substrate solution diluted 1:2 with water (*see* **Note 9**).

9. Expose the membranes to autoradiography films or to a gel imager (*see* example results in Fig. 3) (*see* **Note 10**).

Fig. 2 Schematic representation of the sandwich assembly for the Western blot

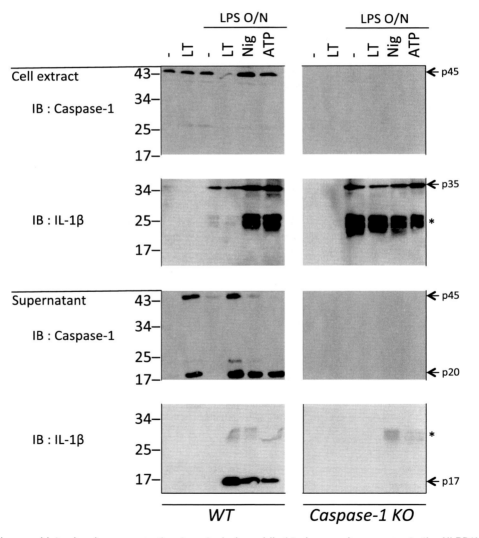

Fig. 3 Immunoblots showing caspase-1 autoproteolysis and IL-1β cleavage in response to the NLRP1b and NLRP3 inflammasome activation in WT and *caspase-1* KO peritoneal macrophages. Caspase-1 and IL-1β are shown in cell extracts and cell supernatants. *O/N* overnight, *LT* lethal toxin, *Nig* nigericin, – untreated, * unspecific signal

3.6   LDH Assay

This assay is carried out following the manufacturer's instructions. It is used to detect the cell death induced by caspase-1 activation.

1. Make sure to set up the experimental controls: 1 well left untreated for cell spontaneous LDH release and 1 well treated with a lysis solution for cell maximum LDH release.

2. Transfer 50 µl of cell supernatant to a flat-bottom 96-well plate.

3. Add 50 µl of the reconstituted LDH Substrate Mix to each well of the plate, cover and incubate at room temperature, protected from light, up to 30 min.

4. Add 50 µl of the Stop Solution to each well of the plate.

5. Read the absorbance at 490 nm.

LDH release percentage is calculated by the following formula:

$$([\text{LDH treated}]-[\text{LDH untreated}]/[\text{LDH total lysis}]-[\text{LDH untreated}])\times100$$

4   Notes

1. Make small aliquots of ATP stock and always use a new aliquot for inflammasome stimulation.

2. Use 12 N HCl to get close to the desired pH then use 6 N HCl to finish the adjustment.

3. Phenol red can be replaced by bromophenol blue. We find convenient, for instance, to use a blue LB for cell extracts and a red LB for cell supernatants.

4. Opti-MEM® medium allows a cleaner protein electrophoresis as it contains very few proteins.

5. It is also possible to resuspend the protein pellet in 30 µl of RIPA buffer 1× and then add LB.

6. The casted gels can be stored for 10 days immerged in a box containing running 1× buffer at +4 °C.

7. If many wells are empty, we recommend adding 15 µl of LB 3× to these empty wells to ensure an even migration of the samples.

8. As the best transfer is obtained with cool buffer, do not remove buffer from the fridge until immediately prior to filling tanks. Some people of the team like to transfer in the cold room.

9. We use the Luminata™ Crescendo Western HRP substrate (Millipore). This HRP substrate is convenient as it is premixed. Furthermore, depending on the abundance of the protein detected and on the specificity of the primary antibody, it is possible to adapt the substrate concentration by diluting it

with water and to also adapt the time of membrane incubation to modulate the intensity of the signal.

10. The membranes can be re-hybridized with another antibody to detect another protein without being stripped if, the size of the other protein of interest to be detected is different and if the primary antibody species is different. To that purpose, we block the membranes as described in Subheading 3.5 and add 0.05 % of sodium azide to the buffer to inactive the HRP. For instance, the anti-caspase-1 antibody can be re-hybridized following the hybridization with the anti-IL-1β antibody.

Acknowledgement

We thank Sarah Karabani for proofreading the manuscript. V. Petrilli is supported by the ITMO cancer and the Ligue contre le cancer comité de l'Ain.

References

1. Walker NP et al (1994) Crystal structure of the cysteine protease interleukin-1 beta-converting enzyme: a (p20/p10)2 homodimer. Cell 78:343–352

2. Elliott JM, Rouge L, Wiesmann C, Scheer JM (2009) Crystal structure of procaspase-1 zymogen domain reveals insight into inflammatory caspase autoactivation. J Biol Chem 284:6546–6553

3. Lamkanfi M, Dixit VM (2012) Inflammasomes and their roles in health and disease. Annu Rev Cell Dev Biol 28:137–161

4. Broz P, von Moltke J, Jones JW, Vance RE, Monack DM (2010) Differential requirement for Caspase-1 autoproteolysis in pathogen-induced cell death and cytokine processing. Cell Host Microbe 8:471–483

5. Guey B, Bodnar M, Manié SN, Tardivel A, Petrilli V (2014) Caspase-1 autopro-teolysis is differentially required for NLRP1b and NLRP3 inflammasome function. Proc Natl Acad Sci U S A 111:17254–17259

Chapter 14

Cell-Free Assay for Inflammasome Activation

Yvan Jamilloux and Fabio Martinon

Abstract

Inflammasomes are multiprotein complexes, which assembly results in caspase-1 activation and subsequent IL-1β and IL-18 activation and secretion. In a cell-free system, based on cytosols of normally growing cells, the disruption of the cell membrane spontaneously activates the inflammasome. Studying the activation of the inflammasome in cytosolic extracts provides multiple advantages, as it is synchronized, rapid, strong, and mostly plasma membrane-free. This protocol covers the methods required to prepare cell lysates and study inflammasome activation using different read-outs. General considerations are provided that may help in the design of modified methods. This assay can be useful to study potential inflammasome interactors and the signaling pathways involved in its activation.

Key words Cell-free assay, Inflammasome, ASC, NLRP3, IL-1β

1 Introduction

Inflammasomes are multiprotein complexes, which are typically composed of a cytoplasmic sensor and an effector protease, caspase-1, which constitutes the catalytic active core of the complex. Most inflammasomes also require the adaptor ASC (apoptosis-associated speck-like protein containing a CARD) that recruits caspase-1 to the oligomeric sensor [1, 2]. The assembly of these complexes depends on the recognition of infectious stimuli or danger signals and results in caspase-1 activation and subsequent IL-1β activation and secretion. The first characterization of an inflammasome complex was made in 2002 in a cell-free system using THP-1 cell extracts [1]. This assay, which is based on the hypotonic lysis of cells in low-potassium buffers (<70 mmol/L), had been adapted from previous studies that characterized and studied the apoptosome [3, 4]. The apoptosome is formed upon binding of cytochrome c to APAF-1, leading to its oligomerization and further recruitment and activation of the apoptotic protease, caspase-9. While the assembly and activation of the apoptosome in cell-free extracts requires the addition of exogenous cytochrome c or its release

Francesco Di Virgilio and Pablo Pelegrín (eds.), *NLR Proteins: Methods and Protocols*, Methods in Molecular Biology, vol. 1417, DOI 10.1007/978-1-4939-3566-6_14, © Springer Science+Business Media New York 2016

from damaged mitochondria, the activation of the inflammasome occurs spontaneously, suggesting that its ligand or trigger is activated or released by mechanical disruption of the cell integrity.

Because the study of the inflammasome using the cell-free assay requires a relatively large amount of cells, the use of immortalized cells facilitates the setup of this assay. However, in most immortalized cell lines the ASC promoter is highly methylated and the transcription of ASC is repressed [5]. One exception is the human monocytic cell line, THP-1, that expresses ASC and is therefore, among human cell lines, the one that best responds to inflammasome activation, including in cell-free conditions [1]. Interestingly, caspase-1 was first detected and purified from THP-1 lysates prepared in buffers containing low-potassium, years before the first description of the inflammasome [6–9]. It is therefore likely that the preparation of the cell extracts in these landmark papers led to the spontaneous assembly and activation of the inflammasome that resulted in the observed caspase-1 activity. Indeed THP-1 cells do not activate caspase-1 or secrete processed IL-1β at basal conditions.

Over the years, the cell-free assay has been used in a few studies that investigated some aspects of the inflammasome. For example, it was used to demonstrate the key role of ASC in THP-1 cells as well as in primary mouse macrophages [1, 10]. It was then used to study the rates of caspase-1 activation in THP-1 cells and Bac1 mouse macrophages [11, 12]. This system has also been used to demonstrate the role of low-potassium concentration on inflammasome activation [13].

This system remains of interest for the study of inflammasome activation and regulation since it allows: (1) rapid, synchronized, and strong detection of inflammasome activation, caspase-1 cleavage, and IL-1β processing; (2) easy manipulation of the system with the addition of putative activators or inhibitors in the cell lysates; and (3) analysis of the biochemistry of the complex including its composition at definite time points following activation, without need for further membrane crossing.

Here, we describe the successive steps required to activate the inflammasome in cell-free extracts obtained from THP-1 monocytes and give examples of available read-outs.

2 Materials

Solutions are to be prepared using ultrapure and deionized water. All reagents, including solutions, tubes, and syringes, should be brought to a temperature of 4 °C unless indicated otherwise.

1. 22G needles.
2. 1 mL syringes.

3. Pasteur pipettes.

4. Thermoshaker block for microtubes, as the ThermoMixer® (Eppendorf) or equivalent such as the Thermo-Shaker® (Biosan) or the ThermoCell® Mixing Block (Bioer).

5. THP-1 human monocytic cell line. (THP-1 cells are cultured to a density of about 1.5×10^6 cells/mL.)

6. 150-cm² plastic flasks for cell culture or, alternatively, 2-L roller bottles system for cell culture.

7. Sterile 50 mL conical tubes.

8. 1.5 mL microfuge tube.

9. Refrigerated centrifuge and microfuge.

10. Complete cell culture media: Roswell Park Memorial Institute (RPMI) 1640 media complemented with 10 % heat-inactivated fetal calf serum (FCS) and 100 µg/mL each of penicillin/streptomycin (*see* **Note 1**).

11. Sterile phosphate buffered saline solution (PBS): 137 mM NaCl, 2.7 mM KCl, 1 mM Na_2HPO_4, 2 mM KH_2PO_4, pH 7.4.

12. Buffer W: 20 mM HEPES–KOH at pH 7.5, 10 mM KCl, 1.5 mM $MgCl_2$, 1 mM sodium EDTA, 1 mM sodium EGTA, and 0.1 mM PMSF (*see* **Note 2**). Since PMSF is unstable in water, it should be added freshly for each experiment (PMSF can be stored for longer time in isopropanol or ethanol). Store Buffer W at 4 °C (*see* **Note 3**).

13. Buffer A: 20 mM HEPES–KOH at pH 7.5, 10 mM KCl, 1.5 mM $MgCl_2$, 1 mM sodium EDTA, 1 mM sodium EGTA, and 320 mM sucrose.

14. CHAPS buffer: 20 mM HEPES–KOH at pH 7.5, 5 mM $MgCl_2$, 0.5 mM sodium EGTA, 0.1 mM PMSF, and 0.1 % CHAPS. PMSF should be added freshly for each experiment.

15. Disuccinimidyl suberate (DSS), prepare fresh each time in dimethyl sulfoxide (DMSO) according to the manufacturer's instructions.

16. Antibody dilution buffer: 5 % skimmed milk.

17. 1:1000 dilution of each primary antibody: mouse monoclonal antihuman caspase-1 (#AG-20B-0048-C100, AdipoGen), rabbit polyclonal anti-ASC (AL177, #AG-25B-0006-C100, AdipoGen), and mouse monoclonal antihuman IL-1β (MAB201, R&D Systems).

18. 1:10,000 dilution of appropriate horseradish peroxidase (HRP)-conjugated secondary antibodies.

19. Nitrocellulose membranes.

20. 1 M DTT stock solution: dissolve 1.5 g of DTT in 8 mL of water, adjust the final volume to 10 mL with water, and

dispense into 1 mL aliquots. Store the stock solution in the dark, at −20 °C.

21. 4× Laemmli Sample Buffer: for 100 mL mix 3 g of Tris–HCl adjust to pH 6.8, 40 mL of a 40 % solution of glycerol, 5 g of 5 % SDS, and 5 mg of 0.005 % bromophenol blue. Complete with water to 100 mL. Store at room temperature and before use complement with 400 mM DTT from stock solution.

22. 15 % SDS-polyacrylamide resolving gel: for 8 mL mix 2.8 mL of water, 3 mL of 40 % acrylamide/bis-acrylamide solution, 2 mL of 1.5 M Tris–HCl pH 8.8, 80 μL of 10 % SDS, 80 μL of 10 % ammonium persulfate (APS), and 8 μL of TEMED immediately before pouring the gel.

23. Stacking SDS-polyacrylamide gel: for 5 mL mix 2.9 mL of water, 750 μL of 40 % acrylamide/bis-acrylamide solution, 1.25 mL of 0.5 M Tris–HCl pH 6.8, 50 μL of 10 % SDS, 50 μL of 10 % APS, and 5 μL of TEMED immediately before pouring the stacking gel.

24. Acrylamide electrophoresis system.

25. Semidry or wet transfer system.

3   Methods

3.1   Cell-Free Lysates Preparation and Inflammasome Activation

Carry out all procedures on ice unless otherwise specified. A schematic diagram of the cell-free assay is provided in Fig. 1.

1. Harvest 100 million of THP-1 cells (*see* **Note 4**) in 50 mL tubes and centrifuge the cells at 4 °C for 5 min at $400 \times g$. Wash the pellets by aspirating the supernatant, adding 2 mL of ice-cold PBS, and centrifuging the tubes again 5 min at $400 \times g$. Discard the supernatant, centrifuge the pellets one more time

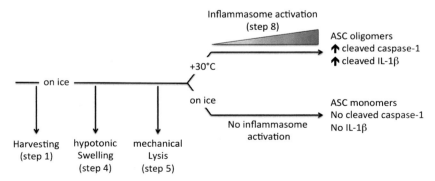

Fig. 1 Schematic diagram of the cell-free assay. After hypotonic swelling and mechanical disruption of the plasmatic membrane, the inflammasome complex is spontaneously activated in the lysates of THP-1 cells, upon incubation at 30 °C

for 2 min at $400 \times g$ and remove the remaining liquid with a Pasteur pipette (*see* **Note 5**).

2. Add 2 mL of ice-cold buffer W, without resuspending cells. Centrifuge the tubes for 5 min at $400 \times g$. Discard the supernatant.

3. Estimate the volume of the dry pellet (*see* **Note 6**) and add $3 \times$ that volume of ice-cold buffer W (*see* **Note 7**). Resuspend the cells by gentle pipetting up and down. Transfer the resuspended cells to a prechilled 1.5 mL microfuge tube. Keep cells on ice.

4. Incubate 10 min on ice to allow the swelling of the cells.

5. Mechanically disrupt cell membranes by 15 passages (*see* **Note 8**) through a 22G needle, with the 1 mL syringe. Work rapidly and stay on ice to avoid the activation of the inflammasome and avoid the generation of air bubbles.

6. Centrifuge the tubes 5 min at $13,000 \times g$, at 4 °C.

7. Harvest the supernatant in a prechilled 1.5 mL microfuge tube. The negative control tube is to be left on ice or at −20 °C. At this time, protein concentration can be checked and adjusted using commercial kits (*see* **Note 9**).

8. Aliquot the samples in experimental tubes using at least 25 μL of lysate per condition (*see* **Note 10**).

9. Incubate the sample tubes during 30–120 min at 30–37 °C (*see* **Note 11**), with agitation set at 400 rpm, in a thermoshaker block (*see* **Note 12**).

10. The process can be stopped at any time, either by freezing the lysates at −20 °C or by adding denaturing buffers (Laemmli sample buffer or buffer A) (*see* **Note 13**).

3.2 Read-Outs

3.2.1 Caspase-1 Cleavage and Pro-IL1β Processing

1. Dilute 4× Laemmli sample buffer (containing 400 mM DTT) to 1× with the samples and the negative control from **step 10** of Subheading 3.1.

2. Boil samples for 5 min.

3. Fractionate samples by electrophoresis on a 15 % SDS-polyacrylamide gel.

4. Transfer proteins into a nitrocellulose membrane.

5. Probe membranes with antibodies for caspase-1 and IL-1β. Processing of these two proteins can then be analyzed, pro-caspase-1 cleavage results in 20 kDa (p20) fragment of active caspase-1 (Fig. 2), and pro-IL-1β processing results in 17 kDa (p17) active fragment of IL-1β.

3.2.2 ASC Assembly Assay

1. Save a 10 % aliquot of each incubated sample from **step 10** of Subheading 3.1 for future Western blot analysis of total ASC.

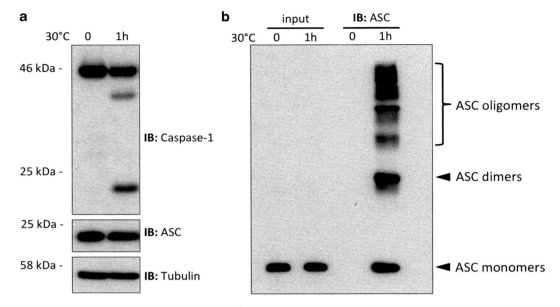

Fig. 2 Possible read-outs for inflammasome activation, using the cell-free assay. (**a**) In THP-1 cell lysates, obtained after hypotonic swelling and mechanical disruption, the inflammasome was activated upon incubation at 30 °C during 1 h, resulting in caspase-1 cleavage (20 kDa band). (**b**) Inflammasome activation can also be demonstrated by the formation of ASC aggregates (oligomers) upon incubation at 30 °C during 1 h

2. In each sample, add twice the sample volume of buffer A and centrifuge the lysates at $2000 \times g$ for 5 min at 4 °C.

3. Collect the supernatant in a clean 1.5 mL microfuge tube. Dilute the supernatant with 1 volume of CHAPS buffer and centrifuge at $5000 \times g$ for 8 min to pellet the ASC oligomers. Discard the supernatant.

4. Wash the ASC aggregates by resuspending the crude pellets in 1 mL of CHAPS buffer, pipet up and down, and centrifuge the tubes for 8 min at $5000 \times g$. This step should be repeated twice.

5. After three washes, discard the supernatant and resuspend the pellets in 30 μL of CHAPS buffer.

6. Crosslink by adding 4 mM DSS and incubate 30 min at room temperature.

7. Quench the reaction with 30 μL of 2× Laemmli buffer without DTT. Add 2× Laemmli buffer without DTT to the 10 % aliquot from **step 1**.

8. The ASC assembly can then be analyzed by Western blot after boiling and fractionating on a 15 % SDS-polyacrylamide gel and using anti-ASC antibody. The assembly of aggregated ASC is detected as monomers (22 kDa), together with dimers, and high molecular weight oligomers (Fig. 2).

4   Notes

1. Other cells that express the inflammasome components (including ASC) can be used. For example, this assay has been successfully reported in thioglycolate-elicited mouse peritoneal macrophages [10], in mouse Bac1 macrophages [11], and in J2-immortalized mouse bone marrow-derived macrophages [[13]; *unpublished personal data*].

2. Buffer W refers to buffer A reported in Xiaodong Wang's landmark studies characterizing the apoptosome complex in a cell-free system [4, 14, 15].

3. A buffer W with 150 mM KCl can be used as negative control, since inflammasome is only activated with potassium concentration below 70 mmol/L. However because the addition of potassium may affect the hypotonic lysis, it is recommended to add the extra KCl after mechanical lysis of the cells (**step 5** of Subheading 3.1).

4. This amount of cells is optimized for this protocol to obtain a good lysis and a reproducible caspase-1 activity, however the assay can be scaled-up or down.

5. Remaining PBS and media may affect the efficiency of the hypotonic lysis.

6. To estimate the volume of the dry pellet, place an empty tube next to the tube containing the pellet, add the same volume of a solution (water) to reach the level of the pellet. Measure the volume of the solution with a pipette.

7. The volume of buffer W can be increased up to 5× the volume of the dry pellet. This may decrease the activity of the lysate but may facilitate downstream applications.

8. The number of passages can vary from 10 to 20, depending on the volume to lyse and the experimenter.

9. For example, the Quick Start™ Bradford Protein Assay (Bio-Rad) can be used. Use the "standard protocol" with 10–100-times dilutions of each sample in buffer W. Bovine serum albumin is used for the standard range. Pipet 5 μL of each sample into separate microplate wells. Dispense 250 μL of 1× dye reagent in each well with multichannel pipet. Depress the plunger repeatedly to mix the sample and reagent in the wells. Incubate 5 min at room temperature. Read the microplate using a spectrophotometer with wavelength set at 595 nm. Refer to manufacturer's protocol for data analysis.

10. We typically performed experiments using 25–500 μL of lysate per condition without observing obvious differences in the activity per μL. The protein concentration of the lysates typically ranges from 5 to 10 μg/μL.

11. In THP-1 lysates the activation occurs within minutes. A 30 °C incubation temperature should be preferred to limit protein degradation. Note that the assay can be performed at 37 °C, this may result in stronger activity depending on the cell line.

12. Strong activation can also be observed using a water bath incubator rather than with the thermoshaker block.

13. Laemmli sample buffer can be used if the samples are directly loaded on a polyacrylamide gel electrophoresis for Western blotting. Buffer A is used if the samples are analyzed for ASC oligomerization. Additionally, RIPA buffer can be used in other biochemical procedures such as immunoprecipitations and molecular pull-down assays.

Acknowledgments

F.M. is supported by a grant from the European Research Council (starting grant 281996), a Human Frontier Science Program career development award (CDA00059/2011), and a grant from the Swiss National Science Foundation (31003A-130476).

Y.J. is supported by a grant from the Foundation for the Development of Internal Medicine in Europe (FDIME), a grant from the *Société Nationale Française de Médecine Interne* (SNFMI)/Genzyme, a grant from Groupama foundation, and a "*poste d'accueil*" at INSERM.

References

1. Martinon F, Burns K, Tschopp J (2002) The inflammasome: a molecular platform triggering activation of inflammatory caspases and processing of proIL-beta. Mol Cell 10:417–426

2. Martinon F, Mayor A, Tschopp J (2009) The inflammasomes: guardians of the body. Annu Rev Immunol 27:229–265

3. Cain K, Bratton SB, Langlais C et al (2000) Apaf-1 oligomerizes into biologically active approximately 700-kDa and inactive approximately 1.4-MDa apoptosome complexes. J Biol Chem 275:6067–6070

4. Salminen A, Kauppinen A, Hiltunen M et al (2014) Epigenetic regulation of ASC/TMS1 expression: potential role in apoptosis and inflammasome function. Cell Mole Life Sci 71:1855–1864

5. Ayala JM, Yamin TT, Egger LA et al (1994) IL-1 beta-converting enzyme is present in monocytic cells as an inactive 45-kDa precursor. J Immunol 153:2592–2599

6. Cerretti DP, Kozlosky CJ, Mosley B et al (1992) Molecular cloning of the interleukin-1 beta converting enzyme. Science 256:97–100

7. Kostura MJ, Tocci MJ, Limjuco G et al (1989) Identification of a monocyte specific pre-interleukin 1 beta convertase activity. Proc Natl Acad Sci U S A 86:5227–5231

8. Miller DK, Ayala JM, Egger LA et al (1993) Purification and characterization of active human interleukin-1 beta-converting enzyme from THP.1 monocytic cells. J Biol Chem 268:18062–18069

9. Yamamoto M, Yaginuma K, Tsutsui H et al (2004) ASC is essential for LPS-induced activation of procaspase-1 independently of TLR-associated signal adaptor molecules. Genes Cells 9:1055–1067

10. Kahlenberg JM, Dubyak GR (2004) Differing caspase-1 activation states in monocyte versus macrophage models of IL-1beta processing and release. J Leukoc Biol 76:676–684

11. Kahlenberg JM, Dubyak GR (2004) Mechanisms of caspase-1 activation by P2X7 receptor-mediated K+ release. Am J Physiol Cell Physiol 286:C1100–C1108

12. Pétrilli V, Papin S, Dostert C et al (2007) Activation of the NALP3 inflammasome is triggered by low intracellular potassium concentration. Cell Death Differ 14:1583–1589

13. Pirami L, Stockinger B, Corradin SB et al (1991) Mouse macrophage clones immortalized by retroviruses are functionally heterogeneous. Proc Natl Acad Sci U S A 88: 7543–7547

14. Zou H, Henzel WJ, Liu X et al (1997) Apaf-1, a human protein homologous to C. elegans CED-4, participates in cytochrome c-dependent activation of caspase-3. Cell 90: 405–413

15. Li P, Nijhawan D, Budihardjo I et al (1997) Cytochrome c and dATP-dependent formation of Apaf-1/caspase-9 complex initiates an apoptotic protease cascade. Cell 91:479–489

Chapter 15

Functional Reconstruction of NLRs in HEK293 Cells

Vincent Compan and Gloria López-Castejón

Abstract

Inflammasomes are molecular complexes that initiate innate immune response. They are mainly expressed by immune cells; however, molecular manipulations in these cells remain very difficult. Here, we describe a simple protocol to overexpress and activate functional NRLP3 inflammasomes in HEK293 cells.

Key words NLR, HEK293, ASC, Inflammasome, Active caspase-1

1 Introduction

Members of the NOD-like receptor (NLR) family are key components in the initiation of innate immune response to tissue injury or pathogen infection. These proteins trigger cellular responses by forming supramolecular complexes called inflammasomes [1]. The best characterized inflammasome contains the NLR protein NLRP3, the adaptor protein ASC, and the cysteine protease caspase-1. Activation of this complex leads to the processing and the release of the pro-inflammatory cytokine IL-1β in order to initiate and propagate the inflammatory response. NLRP3 inflammasome responds to a broad range of stimuli including different pathogens or pathogen associated toxin (as the bacterial ionophore nigericin [2]), endogenous danger signals (as extracellular ATP [2]), or variation in homeostatic parameter (for example, variation in extracellular osmolarity [3]). Immune cells are the main cells expressing the NLRP3 inflammasome, including macrophages and monocytes [4]. Characterization of inflammasome activation remains rather challenging as immune cells are hard to transfect cells and they express endogenous inflammasome components that could interact with the variations under study. Thus, we developed a simple protocol for expression of NLRP3 components in HEK293 cells that allows reconstituting a functional NLR complex. We found that in response to low osmolarity or nigericin, overexpressed NLRP3 inflammasome trigger caspase-1 activation and IL-1β

Francesco Di Virgilio and Pablo Pelegrín (eds.), *NLR Proteins: Methods and Protocols*, Methods in Molecular Biology, vol. 1417, DOI 10.1007/978-1-4939-3566-6_15, © Springer Science+Business Media New York 2016

processing with similar features to an endogenous inflammasome, hence validating the use of this approach in experimental research settings [5].

2 Materials

2.1 Transfection of NLRP3 Inflammasome in HEK293 Cells

1. Lipofectamine2000 reagent® (Life Technologies) or equivalent can be used to transfect HEK293 cells.

2. Complete cell culture media: Dulbecco's modified Eagle's medium (DMEM) high glucose with glutamine and complemented with 10 % decomplemented fetal bovine serum. Antibiotics as penicillin/streptomycin can be used.

3. Opti-MEM® medium (Life Technologies).

4. Mammalian expressing plasmids coding for NLRP3, caspase-1, and ASC (*see* **Note 1**).

2.2 Detection of Inflammasome Functionality/ Activation by Western Blot

1. Antibodies for detection of caspase-1 (inactive form and one of the active forms p10 or p20 of caspase-1), ASC, and NLRP3 (*see* **Note 2**).

2. Cell lysis buffer (CLB): 20 mM HEPES (pH 7.4), 100 mM NaCl, 5 mM EDTA, 1 % Nonidet P-40, and supplemented on day of use with protease inhibitors.

3. HBS solution: 147 mM NaCl, 2 mM KCl, 2 mM $CaCl_2$, 1 mM $MgCl_2$, 10 mM HEPES, and 13 mM d-glucose (pH 7.4).

4. Hypotonic solution: HBS solution diluted in sterile ultrapure H_2O to get a 90 mOsmol solution.

5. Tris–glycine gels (resolving gel 15 % and stacking gel 5 %, *see* **Note 3**).

6. SDS-PAGE running buffer (TGS; 1×): 25 mM Tris–HCl, 192 mM Glycine, 0.1 % (w/v) SDS (pH 8.3).

7. Laemmli Sample Buffer 4× supplemented with 10 % β-mercaptoethanol.

8. Mini-PROTEAN Tetra Cell electrophoresis system (Bio-Rad) or equivalent.

9. Trans-Blot Turbo transfer system (Bio-Rad) or equivalent.

10. Nitrocellulose membrane pore size 0.22 μm.

11. Transfer buffer, for 1 L: 200 mL of 5× Transfer buffer (Bio-Rad), 600 mL of nanopure water, and 200 mL of ethanol 100 %.

12. Powdered skim milk.

13. Bovine serum albumin (BSA).

14. Tris–HCl buffered saline (TBS; 10×): 1.5 M NaCl, 0.1 M Tris–HCl (pH 7.4).

15. TTBS solution: TBS containing 0.1 % Tween-20.

16. Blocking solution: 5 % milk in TBST. Store at 4 °C.

17. Secondary immunoglobulins coupled to HRP.

18. Enhance chemoluminescent (ECL) substrate for Western blot.

19. ChemiDoc MP Imaging System (Bio-Rad) or equivalent.

3   Methods

Carry out all procedures with sterile and pyrogen-free material in biological safety cabinets Class II at room temperature unless otherwise specified.

3.1   Transfection of NLRP3 Inflammasome in HEK293

1. Plate HEK293 on 35 mm dishes to get 60 % confluence. Use one dish per condition to be tested.

2. Transfect cells 4–5 h later using Lipofectamine 2000® manufacturer's instructions with few exceptions. For one 35 mm dish, in a first tube prepare 0.1 μg of each DNA coding for NLRP3 and caspase-1, and 0.2 μg DNA coding for ASC in 100 μL Opti-MEM. In a second tube, gently mix 2.0 μL Lipofectamine 2000® transfection reagent in 100 μL Opti-MEM. Incubate for 5 min at room temperature then add tube 1 into tube 2. After 20 min at room temperature, add the previous mix (DNA-transfection reagent complexes) on dish containing 1 mL of complete cell culture media.

3. The following day, remove transfection reagent containing media and replace by 2 mL of fresh cell media.

3.2   Inflammasome Activation

1. 48 h after transfection, cells expressing functional NLRP3 inflammasome can be used (see **Note 4**).

2. Wash cells twice with HBS.

3. Add 1 ml of HBS to each dish.

4. To activate the inflammasome add 5 μM nigericin for 30 min or apply the hypotonic solution (1 mL) for 40 min.

5. Remove supernatants, add 100 μL of CLB to the cells.

6. Scrape cells and transfer cell lysate to a tube. Store on ice.

7. After 30 min on ice, spin tubes at $13,000 \times g$ at 4 °C to remove insolubilized materials and keep supernatant.

3.3   Western Blot to Detect Inflammasome Functionality/ Activation

1. Determine protein concentration using a BCA kit (alternative protein concentration methods can be used).

2. Load 30 μg of protein form each condition per lane.

3. Run the SDS-PAGE gels and transfer proteins on nitrocellulose membrane (see **Note 5**).

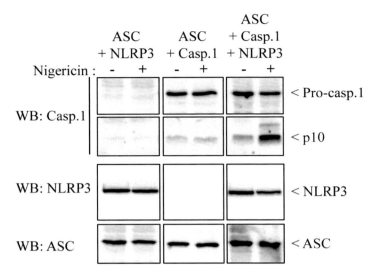

Fig. 1 Reconstruction of functional NLRP3 inflammasome. HEK293 cells overexpressing inflammasome components (NLRP3, Caspase-1, and ASC) were stimulated with the K$^+$ ionophore nigericin (5 µM for 40 min). Activation of caspase-1 (detection of p10 subunit) and expression of NLRP3 and ASC were detected by WB after protein separation on 4–12 % gradient gels. Note that in this experiment, cells were also transfected with the IL-1β BRET sensor described in Chap. 5 (not shown) (adapted from [5])

4. Block the membrane for 30 min in 5 % milk/TBST.

5. Probe membranes with caspase-1, NLRP3, and ASC antibodies accordingly, overnight at 4 °C.

6. Wash the membrane 3× with TBST during 30 min.

7. Incubate membrane with appropriate secondary antibodies in 5 % milk/TBST for 1 h at room temperature.

8. Wash the membrane 3× with TBST during 30 min.

9. Develop using ECL substrate according to manufacturer instructions.

10. Check for the active form of caspase-1 p10 or p20 to determine inflammasome activation. Check for the presence of NLRP3 and ASC to confirm their expression (Fig. 1).

4   Notes

1. Similar protocol described here has been used to reconstitute functional NLRPC4 inflammasome (data not shown).

2. References and dilution of antibodies working by WB: caspase-1 (Santa Cruz, sc-515, dilution 1:300), ASC (Santa Cruz sc-22514-R, dilution 1:1000), and NLRP3 (Adipogen, AG-20B-0014-C100, dilution 1:1000).

3. Alternatively to the current protocol describe here, NLRP3 inflammasome activation can be detected in 96-well plate using the IL-1β BRET biosensor described in this book (*see* Chap. 5).

4. 15 % acrylamide homemade gels can be employed to visualize active caspase-1. Alternatively, 4–12 % precast gradient gels allow large separation of proteins with different molecular weight as NLRP3 (MW = 118 kDa) and p10 fragment of active caspase-1 (MW = 10 kDa).

5. To improve caspase-1 p10 detection, transfer proteins on nitrocellulose membrane with pore size of 0.22 μm.

Acknowledgments

This work was supported by funds from the Manchester Collaborative Centre for Inflammation Research and a Sir Henry Dale fellowship from the Wellcome Trust (UK) to G.L.-C. V.C. was supported by the *Institut National de la Santé et de la Recherche Médicale.*

References

1. Martinon F, Burns K, Tschopp J (2002) The inflammasome: a molecular platform triggering activation of inflammatory caspases and processing of proIL-beta. Mol Cell 10(2):417–426

2. Mariathasan S, Weiss DS, Newton K, McBride J, O'Rourke K, Roose-Girma M, Lee WP, Weinrauch Y, Monack DM, Dixit VM (2006) Cryopyrin activates the inflammasome in response to toxins and ATP. Nature 440(7081):228–232. doi:10.1038/nature04515

3. Compan V, Baroja-Mazo A, Lopez-Castejon G, Gomez AI, Martinez CM, Angosto D, Montero MT, Herranz AS, Bazan E, Reimers D, Mulero V, Pelegrin P (2012) Cell volume regulation modulates NLRP3 inflammasome activation. Immunity 37(3):487–500. doi:10.1016/j.immuni.2012.06.013

4. Guarda G, Zenger M, Yazdi AS, Schroder K, Ferrero I, Menu P, Tardivel A, Mattmann C, Tschopp J (2011) Differential expression of NLRP3 among hematopoietic cells. J Immunol 186(4):2529–2534. doi:10.4049/jimmunol.1002720

5. Compan V, Baroja-Mazo A, Bragg L, Verkhratsky A, Perroy J, Pelegrin P (2012) A genetically encoded IL-1beta bioluminescence resonance energy transfer sensor to monitor inflammasome activity. J Immunol 189(5):2131–2137. doi:10.4049/jimmunol.1201349

Chapter 16

Method to Measure Ubiquitination of NLRs

Pablo Palazón-Riquelme and Gloria López-Castejón

Abstract

Posttranslational modifications are crucial in determining the functions of proteins in the cell. Modification of the NLRP3 inflammasome by the ubiquitin system has recently emerged as a new level of regulation of the inflammasome complex. Here, we describe a method to detect polyubiquitination of NRLP3 using two different approaches: (1) detection with an ubiquin antibody or (2) using TUBE (Tandem Ubiquitin Binding entities). This approach can be used to detect ubiquitination of other NLR or other components of the inflammasome.

Key words Immunoprecipitation, Ubiquitination, Inflammasome, NLRP3, TUBE

1 Introduction

Ubiquitination is a posttranslational protein modification crucial to maintain cellular homeostasis. The addition of ubiquitin to a protein is mediated by an E1-ubiquitin activating enzyme, an E2-ubiquitin-conjugating enzyme, and an E3 ubiquitin ligase [1]. This is a reversible process and this is controlled by a family of enzymes called deubiquitinases [2]. Disturbing this balanced ubiquitination state can have detrimental consequences for the cell impairing important processes that maintain the normal functioning of the cells such as protein degradation, trafficking, or gene expression [3] and hence the body.

The ubiquitin system also plays a crucial role in the regulation of inflammation. This is especially relevant in innate immune responses where key signalling pathways are finely tuned by ubiquitin [4]. Pathways such as the NFkB rely on this posttranslational modification to successfully initiate the transcription of important proinflammatory mediators [5, 6]. Increasing evidencing, however, is showing that ubiquitin is not only important in the regulation of transcriptional pathways but also in the assembly of the NLRP3 inflammasome [7]. This is a molecular complex that mediates the release of the proinflammatory cytokine IL-18 and

Francesco Di Virgilio and Pablo Pelegrín (eds.), *NLR Proteins: Methods and Protocols*, Methods in Molecular Biology, vol. 1417, DOI 10.1007/978-1-4939-3566-6_16, © Springer Science+Business Media New York 2016

IL-1β [8]. Although it has been shown that the NLRP3 is ubiquitinated, how the ubiquitination state of this receptor regulates the inflammasome still remains a mystery. In order to further study these mechanisms, it is necessary to determine the level of ubiquitination of the NLRP3. In this chapter, we described a method to detect endogenous NLRP3 ubiquitination in NLRP3-expressing HEK293 cells by two different methods. Here, NLRP3 is immunoprecipitated and then the ubiquitination state detected by western blot using either a ubiquitin antibody or Tandem Ubiquitin Binding Entities (TUBE). This method could also be applied to detect ubiquitination in cells that endogenously express NLRP3 or for other components of the inflammasome complex.

2   Materials

2.1   Cell Lysis

1. SDS-lysis buffer: 20 mM Tris–base (pH 8.0), 250 mM NaCl, 3 mM EDTA, 10 % glycerol, 1 % SDS, 0.5 % NP-40, 20 mM N-Ethylmaleimide (NEM), and 5 mM 1,10-phenanthroline monohydrate, protease inhibitors [9].

2. Nondenaturing lysis buffer: 50 mM Tris–HCl (pH 7.4), 150 mM NaCl, 1 mM EDTA, 1 % Triton X-100, 10 % glycerol, 20 mM NEM, and 5 mM 1,10-phenanthroline monohydrate, protease inhibitors (see **Note 1**).

2.2   Transfection

1. Complete cell culture media: Dulbecco's Modified Eagle's Medium (DMEM) containing glucose, l-glutamine, sodium pyruvate, and sodium bicarbonate, supplemented with 10 % of decomplemented Fetal Bovine Serum (FBS).

2. Microcentrifuge tubes.

3. 12-well cell culture plates.

4. Transfection media: DMEM without FBS supplementation.

5. Lipofectamine 2000® transfection reagent (Life Technologies). Other transfection reagents might be also suitable if transfection efficiency is high.

6. Plasmid coding for NLRP3-Flag (see **Note 2**).

2.3   Immunoprecipitation

1. SDS-immunoprecipitation (SDS-IP) buffer: 20 mM Tris–Base (pH 8.0), 250 mM NaCl, 3 mM EDTA, 10 % glycerol, 0.1 % SDS, 0.5 % NP-40, 20 mM NEM, and 5 mM 1,10-phenanthroline monohydrate, protease inhibitors (see **Note 3**).

2. Nondenaturing lysis buffer: 50 mM Tris, 150 mM NaCl, 1 mM EDTA, 1 % Triton X-100, 10 % glycerol, 20 mM NEM, and 5 mM 1,10-phenanthroline monohydrate, protease inhibitors (see **Note 4**).

3. Bicinchoninic acid (BCA) protein assay kit.

4. Anti-Flag M2 affinity gel.

5. Protein G Sepharose.

6. Hamilton syringe or equivalent device, such a narrow-end Pasteur pipette.

2.4 Western Blot

1. Tris-glycine gels (resolving gel 6 % and stacking gel 5 %).

2. Tris-Glycine-SDS (TGS) running buffer: 25 mM Tris, 192 mM Glycine, 0.1 % (w/v) SDS, pH 8.3.

3. Laemmli Sample Buffer 4× supplemented with 10 % β-mercaptoethanol.

4. Protein electrophoresis system. We use the Mini-PROTEAN Tetra Cell electrophoresis system (Bio-Rad).

5. Semidry blot transfer system. We use the Trans-Blot Turbo transfer system (Bio-Rad).

6. Nitrocellulose membrane with pore size 0.45 μm.

7. Transfer buffer, for 1 L: 200 ml of 5× Transfer buffer (Bio-Rad), 600 ml of nanopure water, and 200 ml of ethanol 100 %.

8. Powdered skim milk.

9. Bovine serum albumin (BSA).

10. Tris buffered saline (TBS) 10×: 1.5 M NaCl, 0.1 M Tris–HCl, pH 7.4.

11. Phosphate buffered saline (PBS) 10×: 1.2 M Sodium chloride, 90 mM disodium hydrogen orthophosphate, 37 mM sodium dihydrogen orthophosphate, and 26 mM potassium chloride (pH 7.4).

12. PBST solution: PBS containing 0.1 % Tween-20.

13. TTBS solution: TBS containing 0.1 % Tween-20.

14. Blocking solution: 5 % milk in PBST (BS1) or TBST (BS2). Store at 4 °C.

15. Diluent solution: 5 % milk in TBST (TBST1) or PBST (PBST2). Store at 4 °C.

16. Antibodies: Mouse anti-ubiquitin (SantaCruz, sc-8017), mouse anti-ubiquitin M1-specific (Lifesensor, AB130), mouse anti-NLRP3 (Enzo, ALX-804-818), and rabbit anti-Flag (Cell Signaling, 2368).

17. Secondary Polyclonal rabbit antimouse or goat antirabbit immunoglobulins conjugated with HRP.

18. TUBE2-Biotin (Binds to K6-, M1-, K48-, and K63-linked polyubiquitin) (Lifesensors).

19. Streptavidin-HRP.

20. Enhanced chemoluminescence (ECL) substrate.

21. Imaging System, such as the ChemiDoc MP (Bio-Rad).

3 Methods

3.1 Transfection

1. Plate HEK293 cells (0.15×10^6 cells per well in a 12 well plate) in 1 ml of complete DMEM cell culture media (*see* **Note 5**).

2. The following day transfect cells with 1 μg of NLRP3-Flag plasmid. Control cells were transfected with an empty plasmid. Mix 50 μl of transfection media with 1 μl of NLRP3-Flag-tag vector or empty vector (1 μg/ul) (Tube 1). Mix 50 μl of transfection media with 2.5 μl of Lipofectamine2000 (Tube 2). Incubate for 5 min a room temperature. Add contents from tube 1 into tube 2 and mix gently. Incubate at room temperature for 15 min.

3. Add this mix to the well and incubate the cells for 48 h (*see* **Note 6**).

3.2 Cell Lysis for Detection with Ubiquitin Antibody

1. Wash cells in PBS.

2. Add 200 μl of SDS-lysis buffer per well.

3. Boil samples at 90–95 °C by placing the cell culture plate on the heat block for 20–30 min. To help lysis, mix and pipette up and down using a pipette (10–1000 μl range) for 5–10 s every 3 min during the boiling step. Continue procedure until viscosity is eliminated.

4. Transfer samples to a microcentrifuge tube and centrifuge the samples at $12,000 \times g$, 10 min at room temperature in a standard table-top microfuge and transfer supernatants to new tubes. Cell lysates can now be used or stored at −20 °C before further analysis.

3.3 Cell Lysis for Detection with TUBE-Biotin

1. Perform lysis at 4 °C.

2. Lyse cells by adding 200 μl of nondenaturing lysis buffer per well.

3. Incubate on ice for 10 min.

4. Transfer samples to an microcentrifuge tube and centrifuge the samples at $12,000 \times g$, 10 min at 4 °C in a standard table-top microfuge and transfer supernatants to new tubes. Cell lysates can now be used or stored at −20 °C before further analysis.

3.4 Immunoprecipitation of NLRP3

1. Subject equal amounts of protein lysates (around 100–150 μg) to immunoprecipitation (IP) assay. Measure protein concentration using a BCA assay. Save 5–10 μg of cell lysates for use as the pre-IP samples.

2. Dilute lysates into IP buffer to a final 800 μl of and perform all subsequent steps at 4 °C.

3. To prepare the Sepharose and/or anti-Flag M2 affinity gel, thoroughly suspend them to make a uniform suspension of the resin. Immediately transfer 40 μl of the suspension per reaction to a fresh test tube (see **Note 7**).

4. Wash the resin with IP buffer, 6 times with 1 ml of the corresponding IP buffer. Centrifuge at $8000 \times g$ for 30 s and discard buffer after each wash. In order to let the resin settle in the tube, wait for 1–2 min before removing the buffer (see **Notes 8** and **9**).

5. Preclearing step: Add the 800 μl of cell lysate to the washed resin (Protein G sepharose).

6. Incubate at 4 °C for 1 h with rotation (a roller shaker is recommended). Centrifuge at $8000 \times g$ for 30 s, collect supernatant, and discard agarose (see **Notes 10** and **11**).

7. Add this supernatant to the anti-Flag M2 affinity gel previously washed (see **Note 12**).

8. Agitate or shake all samples and controls gently (a roller shaker is recommended) for 4 h at 4 °C.

9. Centrifuge the resin for 30 s at $8000 \times g$. Remove the supernatants with a narrow-end pipette tip.

10. Wash the resin 5 times with 0.5 ml of the appropriate IP buffer depending on protocol (as in **step 4**). Make sure all the supernatant is removed by using a Hamilton syringe or equivalent device.

11. Dilute the SDS-PAGE sample buffer 4× to 1× with water.

12. Add 40 μl of 1× sample buffer to each sample and control.

13. Boil the samples and controls for 3 min.

14. Centrifuge the samples and controls at $8000 \times g$ for 30 s to pellet undissolved agarose. Transfer the supernatants to fresh test tubes with a Hamilton syringe or a narrow-end Pasteur pipette. These samples are ready to be loaded in a SDS-page gel.

3.5 Western Blot Analysis

1. Load samples obtained from IP in two different 6 % Tris-glycine gels (a) to detect NLRP3 ubiquitination (load 30 μl of IP-sample) and (b) to confirm that IP has worked (load 10 μl of IP-sample).

2. Optional: Load a third gel with Pre-IP samples. This will inform us of the success rate of transfection (see **Note 13**).

3. Run the gel at 150 V in TGS buffer for 1 h and transfer into a nitrocellulose membrane using the High Molecular weight settings in the Trans-Blot Turbo transfer system.

3.6 Detection of Polyubiquitination by Using Ubiquitin Antibodies

1. Transfer membrane into BS1 and block for 2 h at room temperature.

2. Add the desired primary antibodies: anti-Ub (1:200), anti-M1 Ub (1:1000), anti-Flag (1:800), or anti-NLRP3 (1:1000), and incubate overnight at 4 °C.

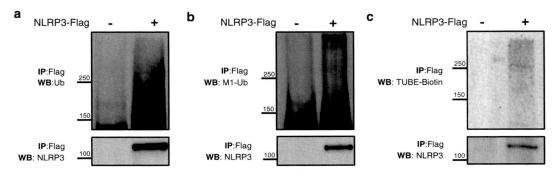

Fig. 1 Western blotting detection of NLRP3 ubiquitination in HEK293 cells. The *blots at the top* show a smear above 120 kDa typical of protein polyubiquitination. This is detected using ubiquitin general antibody (Ub) (**a**), specific linear M1-ubiquitin antibody (M1-Ub) (**b**), or TUBE-Biotin (**c**). The *blots at the bottom* show that the NLRP3 receptor was successfully immunoprecipitated, validating the assay

3. Wash membrane 10×, 2 min each with PBST.

4. Add secondary antibody in BS1 and incubate for 1 h at room temperature.

5. Wash membrane 10×, 2 min each with PBST.

6. Develop using ECL detection reagents according to manufacturer instructions and a luminescence compatible Imaging System (Fig. 1a, b).

3.7 Detection of Polyubiquitination by Using TUBE-Biotin

1. Transfer membrane into BS2 and block for 2 h at room temperature.

2. Add 1 μg/ml of TUBE-Biotin or primary antibodies: anti-Flag (1:800) or anti-NLRP3 (1:1000), and incubate overnight at 4 °C.

3. Wash membrane 10×, 2 min each with PBST.

4. To detect TUBE-Biotin add streptavidin-HRP (1:10,000) in BS2. For other antibodies, add appropriate secondary antibodies in BS1. Incubate for 1 h at room temperature.

5. Wash membrane 10×, 2 min each with PBST.

6. Develop using ECL detection reagents according to manufacturer instructions and a luminescence compatible Imaging System (Fig. 1c).

4 Notes

1. These buffers can be made ahead of time with the exception that the 10 μl/ml of 100× protease inhibitor, NEM, and 1,10-phenanthroline monohydrate should be added immediately before use. The buffers can be stored at room temperature.

2. Plasmid coding for NLRP3 was a gift of J. Tschopp.

3. For detection using ubiquitin antibodies.

4. For detection using TUBE-Biotin.

5. Cells should be 75–80 % confluent when transfecting.

6. Scale the transfection reagents up or down depending on the samples required for your experiment.

7. For resin transfer, use a clean, plastic pipette tip with the end enlarged to allow the resin to be transferred.

8. SDS-IP buffer for ubiquitin and nondenaturing buffer for TUBE protocol.

9. In case of numerous immunoprecipitation samples, wash the resin needed for all samples together and after washing, divide the resin according to the number of samples tested.

10. This step is to remove nonspecific binding to agarose.

11. Remove the supernatant with a narrow-end pipette tip or a Hamilton syringe, being careful not to transfer any resin. Narrow-end pipette tips can be made using forceps to pinch the opening of a plastic pipette tip until it is partially closed.

12. If not using anti-Flag M2 affinity gel, combine the appropriate antibody (e.g., anti-NLRP3 if using endogenous expression) with the cell lysates for 1–4 h at 4 °C to form the immune complexes. Depending on the antibody, different amounts of antibody could be used. Typically, 3–5 μg of antibody is used for each IP. This could be increased to 10 μg if no signal is observed.

13. Dilute them in SDS-load buffer to 1× and boil for 3 min previous to loading.

Acknowledgements

This work was supported by funds from the Manchester Collaborative Centre for Inflammation Research and a Sir Henry Dale fellowship from the Wellcome Trust (UK).

References

1. Hershko A, Ciechanover A (1998) The ubiquitin system. Annu Rev Biochem 67:425–479

2. Clague MJ, Barsukov I, Coulson JM, Liu H, Rigden DJ, Urbe S (2013) Deubiquitylases from genes to organism. Physiol Rev 93:1289–1315

3. Aguilar RC, Wendland B (2003) Ubiquitin: not just for proteasomes anymore. Curr Opin Cell Biol 15:184–190

4. Malynn BA, Ma A (2010) Ubiquitin makes its mark on immune regulation. Immunity 33:843–852

5. Harhaj EW, Dixit VM (2012) Regulation of NF-kappaB by deubiquitinases. Immunol Rev 246:107–124

6. Skaug B, Jiang X, Chen ZJ (2009) The role of ubiquitin in NF-kappaB regulatory pathways. Annu Rev Biochem 78:769–796

7. Lopez-Castejon G (2013) Regulation of NLRP3 activation by the ubiquitin system. Inflammasome 1:15–19

8. Latz E, Xiao TS, Stutz A (2013) Activation and regulation of the inflammasomes. Nat Rev Immunol 13:397–411

9. Hwang J, Kalejta RF (2011) In vivo analysis of protein sumoylation induced by a viral protein: detection of HCMV pp 71-induced Daxx sumoylation. Methods 55:160–165

Chapter 17

Cytofluorometric Quantification of Cell Death Elicited by NLR Proteins

Valentina Sica*, Gwenola Manic*, Guido Kroemer, Ilio Vitale, and Lorenzo Galluzzi

Abstract

Nucleotide-binding domain and leucine-rich repeat containing (NLR) proteins, also known as NOD-like receptors, are critical components of the molecular machinery that senses intracellular danger signals to initiate an innate immune response against invading pathogens or endogenous sources of hazard. The best characterized effect of NLR signaling is the secretion of various cytokines with immunostimulatory effects, including interleukin (IL)-1β and IL-18. Moreover, at least under specific circumstances, NLRs can promote regulated variants of cell death. Here, we detail two protocols for the cytofluorometric quantification of cell death-associated parameters that can be conveniently employed to assess the lethal activity of specific NLRs or their ligands.

Key words Annexin A5, Apoptosis, Caspase, Mitochondria, Necrosis

1 Introduction

Nucleotide-binding domain and leucine-rich repeat containing (NLR) proteins, also known as NOD-like receptors, are a relatively large family of cytosolic, catalytically inactive polypeptides with a key role in the initiation of innate immune responses [1–3]. NLRs have indeed the ability to respond to a wide panel of danger signals by establishing a state of alert that, at least hypothetically, allows for the restoration of homeostasis. Such a response is generally centered on the secretion of proinflammatory cytokines like interleukin (IL)-1β and IL-18, which involves two sequential processes [1, 4]. First, proinflammatory transcription factors including (but not limited to) NF-κB drive the synthesis of the immature forms of IL-1β and IL-18. Second, a supramolecular complex involving caspase-1 catalyzes the proteolytic maturation of IL-1β and IL-18, allowing for their secretion [5, 6]. Of note, NLRs are not only

Valentina Sica and Gwenola Manic both authors contributed equally.

Francesco Di Virgilio and Pablo Pelegrín (eds.), *NLR Proteins: Methods and Protocols*, Methods in Molecular Biology, vol. 1417, DOI 10.1007/978-1-4939-3566-6_17, © Springer Science+Business Media New York 2016

activated by exogenous microbe-associated molecular patterns (MAMPs), i.e., relatively conserved bacterial or viral constituents, but also by endogenous danger-associated molecular patterns (DAMPs) [4, 7, 8]. DAMPs are emitted by eukaryotic cells that experience a situation of stress, alerting neighboring cells of the hazard and playing a central role in the maintenance of organismal homeostasis [9–11]. Interestingly, activated NLRs can also elicit regulated variants of cell death, at least under some circumstances [1]. From a finalistic perspective, this can also be viewed as a mechanism that contributes to the preservation of organismal fitness by ensuring the removal of excessively compromised, and hence potentially dangerous, cells [12, 13].

At odds with its accidental counterparts, regulated cell death (RCD) near-to-invariably occurs when adaptive responses to homeostatic perturbations fail [12, 14], and it relies on the activation of a specific molecular machinery, implying that its course can be pharmacologically or genetically altered [15–17]. Nonetheless, it is difficult to identify the precise moment at which the cascade of events initiated by the failure of adaptive stress responses becomes irreversible [15]. For this reason, although the loss of plasma membrane barrier functions is a late event of most signal transduction cascades that lead to cell death, occurring when most bioenergetic activities are already irremediably compromised, it is nowadays considered as the most reliable marker of cellular demise [15, 16]. In vitro, plasma membrane permeabilization can be easily monitored with so-called exclusion dyes, i.e., colored or fluorescent dyes that cannot enter cells with normal plasma membrane barrier functions [18, 19]. Propidium iodide (PI) and 4′,6-diamidino-2-phenylindole (DAPI) are commonly employed for the quantification of cells with permeabilized plasma membrane by flow cytometry [18, 19].

Additional events that are widely considered as cornerstones in the transition between a reversible fluctuation in cellular homeostasis and the irrevocable commitment of cells to death are mitochondrial outer membrane permeabilization (MOMP) [20–22], the massive activation of some proteases, notably, caspase-3 [16, 23], and the phosphorylation of mixed lineage kinase domain-like (MLKL) [24–26]. Irrespective of how it is elicited, widespread MOMP entails the structural and functional breakdown of the mitochondrial network, producing a bioenergetic and redox crisis [15]. This is lethal for the cell and is invariably accompanied by the dissipation of the electrochemical gradient built across the inner mitochondrial membrane by respiratory chain complexes, i.e., the mitochondrial transmembrane potential ($\Delta\psi_m$) [27, 28]. Accordingly, MOMP can be reliably quantified in vitro by means of $\Delta\psi_m$-sensitive dyes, such as 3,3′-dihexyloxacarbocyanine iodide ($DiOC_6(3)$) or tetramethylrhodamine methyl ester (TMRM) [18]. Caspase-3 activation, which in some (but not all) instances originates from MOMP, has been considered for a long time as the actual cause of a specific RCD instance known as apoptosis [16, 23]. Although the causal relationship between caspase-3 activation and RCD may be less robust than previously thought, caspase-3 activation and some of its manifestations are still commonly measured as indicators of ongoing RCD [15, 29]. This applies, for instance, to the exposure of phosphatidylserine (PS) on the outer leaflet of the plasma membrane, a caspase-dependent process catalyzed by ATPase, class VI, type 11C (ATP11C) [30]. PS exposure can be conveniently monitored on flow cytometry or fluorescence microscopy by means of fluorescent variants of annexin A5 (ANXA5, best known as AnnV), a naturally occurring PS-binding protein [18, 19]. MLKL phosphorylation has recently been identified as a reliable biomarker of necrotic variants of RCD [31, 32], and can be evaluated on flow cytometry or immunoblotting by neophosphoepitope-specific monoclonal antibodies [31, 32]. The assessment of MLKL phosphorylation, however, has not yet become part of routine procedures for the quantification of cell death.

Here, we detail two methods for the cytofluorometric assessment of cell death manifestations based on PI, $DiOC_6(3)$, and AnnV. These protocols are suitable for the quantification of cells that are dead (i.e., they have undergone plasma membrane permeabilization) or committed to die (i.e., they exhibit widespread MOMP and/or massive caspase-3 activation), in vitro, relatively early (24–96 h) after exposure to a lethal trigger. In particular, these methods can be employed to assess the ability of specific NLRs or their ligands to trigger RCD. Moreover, if appropriate pharmacological modulators, such as the broad spectrum caspase inhibitor N-benzyloxycarbonyl-Val-Ala-Asp(O-Me) fluoromethylketone (Z-VAD-fmk) and the receptor-interacting protein kinase 1 (RIPK1) inhibitor Necrostatin-1 (Nec-1) are involved [33–36], these methods allow for a preliminary characterization of the signal transduction pathways that are activated along with (and perhaps account for) the execution of NLR-elicited RCD.

2   Materials

Unless otherwise specified, fluorochromes were obtained from Molecular Probes®-Life Technologies, Fluorochoromes from alternative providers are also suitable (as long as compatible with equipment).

2.1   Disposables

1. 1.5 mL microcentrifuge tubes.
2. 15 and 50 mL conical centrifuge tubes.
3. 5 mL, 12 × 75 mm FACS tubes.
4. 6-, 12-, 24-well plates for cell culture.
5. 75 or 175 cm² flasks for cell culture.

| 2.2 Equipment | Cytofluorometer, such as the FACScan or FACSVantage (Becton Dickinson), equipped with an argon ion laser emitting at 488 nm and controlled by operational/analytical software (*see* **Note 1**). |

2.3 Cell Culture and Treatments

1. Complete growth medium for HCT 116 cells: McCoy's 5A medium containing 3 g/L D-glucose and 1.50 mM L-glutamine, supplemented with 1 mM sodium pyruvate, 100 mM 2-(4-(2-hydroxyethyl)piperazin-1-yl)ethanesulfonic acid (HEPES) buffer and 10 % fetal calf serum (FCS) (*see* **Note 2**).

2. Phosphate buffered saline (PBS, 1×): 137 mM NaCl, 2.7 mM KCl, 4.3 mM Na_2HPO_4, 1.4 mM KH_2PO_4 in deionized water (dH_2O), adjust pH to 7.4 with 2N NaOH.

3. Trypsin/EDTA: 0.25 % trypsin—0.38 g/L (1 mM) EDTA×4 Na^+ in Hank's balanced salt solution (*see* **Notes 3** and **4**).

4. *cis*-diamminedichloroplatinum(II) (CDDP, cisplatin), 50 mM stock solution in N,N-dimethylformamide (DMF), stored at RT under protection from light (*see* **Notes 5–8**).

5. N-benzyloxycarbonyl-Val-Ala-Asp(O-Me) fluoromethylketone (Z-VAD-fmk), 20 mM stock solution in dimethylsulfoxide (DMSO), stored at –20 °C (*see* **Notes 9–12**).

2.4 $DiOC_6(3)$/PI Co-staining

1. 3,3'-dihexiloxalocarbocyanine iodide ($DiOC_6(3)$), 40 µM stock solution in 100 % ethanol, stored at –20 °C under protection from light (*see* **Notes 13–15**).

2. Propidium iodide (PI), 1 mg/mL stock solution in dH_2O, stored at 4 °C under protection from light (*see* **Notes 16–18**).

2.5 FITC-AnnV/PI Co-staining

1. Fluorescein isothiocyanate-conjugated Annexin V (FITC-AnnV), stored at 4 °C (*see* **Notes 19–21**).

2. PI, 1 mg/mL stock solution in dH_2O, stored at 4 °C under protection from light (*see* **Notes 16–18**).

3. Binding buffer (20×) from Miltenyi Biotec, stored at 4 °C (*see* **Note 22**).

3 Methods

3.1 Cell Culture

1. Upon thawing, HCT 116 cells are maintained in complete growth medium within 75 cm² flasks (37 °C, 5 % CO_2) (*see* **Note 23**).

2. As cells approach 80–90 % confluence (*see* **Note 24**), remove culture medium by aspiration, wash gently the adhering cell monolayer with prewarmed PBS (*see* **Note 25**), and add ~3 mL 0.25 % (w/v) trypsin–EDTA solution (*see* **Note 26**).

3. Incubate cells in trypsin/EDTA solution for 1–3 min at 37 °C and, as soon as they are detached (*see* **Notes 27–29**), add complete medium to the cell suspension (*see* **Note 30**).

4. For the propagation of maintenance cell cultures, transfer aliquots of the cell suspension to new 75 cm² flasks and culture them as described in **steps 2–4** (*see* **Notes 31** and **32**, *see also* **Notes 24–30**).

5. Alternatively, for experimental determinations (*see* **Note 33**), seed cells in 12-well plates at a concentration of 0.8×10^5 cells in 1 mL growth medium per well (*see* **Note 34**).

6. Twenty-four hours after seeding (*see* **Notes 35**), treat cells by substituting the culture medium with complete medium supplemented with the NLR agonist of choice (*see* **Notes 36–38**), alone or in the presence of 50 μM Z-VAD-fmk. CDDP (final concentration = 50 μM) can be conveniently employed as a positive control for the induction of Z-VAD-fmk-sensitive cell death (*see* **Note 39**).

3.2 DiOC₆(3)/PI Co-staining

1. Seed and treat HCT 116 cells as described in Subheading 3.1, **steps 5** and **6** (*see also* **Notes 33–39**).

2. At the end of the incubation period, transfer culture supernatants to 5 mL FACS tubes (*see* **Notes 40** and **41**) and detach adherent cells with ~0.5 mL trypsin/EDTA upon a wash with ~1 mL prewarmed PBS (*see also* Subheading 3.1, **steps 2** and **3**, and **Notes 24–29**).

3. Upon complete detachment, add 1 mL complete growth medium (*see* **Note 30**) and collect cells from each well in the FACS tube containing the corresponding supernatant.

4. Centrifuge cell suspensions at $300 \times g$, RT, for 5 min.

5. Discard supernatants and resuspend cells in 200–400 μL of 40 nM DiOC₆(3) in complete growth medium (staining solution) (*see* **Notes 42–45**).

6. Incubate cells with the staining solution for 20–30 min in the dark at 37 °C (5 % CO_2) (*see* **Note 46**).

7. Add PI to a final concentration of 1 μg/mL and incubate samples for additional 2–5 min under protection from light (*see* **Notes 47** and **48**).

8. Acquire and analyze samples by means of a classic cytofluorometer allowing for the simultaneous quantification of light scattering parameters (forward and side scatter, FSC and SSC) and fluorescence signals in two separate channels (e.g., green and red) (*see* **Notes 49–54**).

9. Representative dot plots and quantitative data are illustrated in Fig. 1.

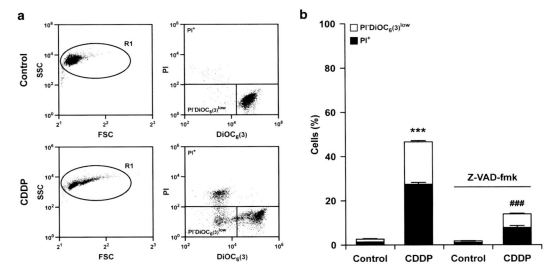

Fig. 1 Cytofluorometric assessment of plasma membrane permeabilization and mitochondrial transmembrane potential dissipation. Human colorectal carcinoma HCT 116 cells were maintained in control conditions or exposed to 50 μM cisplatin (CDDP), alone or combination with 50 μM Z-VAD-fmk for 24 h, then processed for the cytofluorometric quantification of plasma membrane permeabilization and mitochondrial transmembrane potential dissipation as detailed in Subheading 3.2. In *panel a*, representative dot plots are reported. In *panel b*, quantitative data are reported (means ± SD, *n* = 2 parallel samples; ***p < 0.001, as compared to untreated cells; ###p < 0.001, as compared to cells treated with CDDP only; two-sided, unpaired Student's *t* test)

3.3 FITC-AnnV/PI Co-staining

1. Seed, treat, and collect HCT 116 cells as described in Subheading 3.2, **steps 1–4** (*see also* **Notes 24–30** and **33–41**).

2. Gently wash cells 1–2 times with 250–500 μL cold (4 °C) 1× binding buffer (*see* **Note 55**).

3. Prepare the FITC-AnnV staining solution (50–100 μL for tube) by mixing 48–96 μL 1× cold (4 °C) binding buffer and 2–4 μL FITC-AnnV (*see* **Note 56**).

4. Resuspend cells in 50–100 μL FITC-AnnV solution and incubate samples for 15 min at RT under protection from light (*see* **Notes 57** and **58**).

5. Add 1 mL of cold (4 °C) 1× binding buffer and wash 1–2 times.

6. Resuspend cells in 250–500 μL of cold (4 °C) 1× binding buffer previously complemented with 0.5–1 μg/mL PI.

7. Incubate samples for additional 2–5 min at RT, under protection from light, and then proceed to cytofluorometric acquisitions (*see* Subheading 3.2, **step 8**, and **Notes 52–54** and **59**).

8. Representative dot plots and quantitative data are illustrated in Fig. 2.

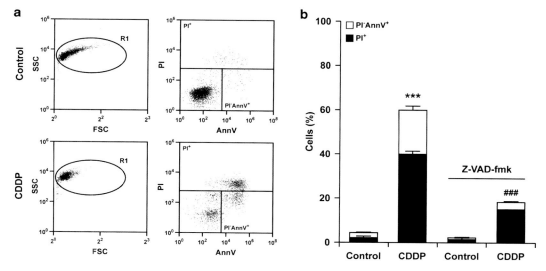

Fig. 2 Cytofluorometric assessment of plasma membrane permeabilization and phosphatidylserine exposure. Human colorectal carcinoma HCT 116 cells were maintained in control conditions or exposed to 50 μM cisplatin (CDDP), alone or combination with 50 μM Z-VAD-fmk for 24 h, then processed for the cytofluorometric quantification of plasma membrane permeabilization and phosphatidylserine exposure as detailed in Subheading 3.3. In *panel a*, representative dot plots are reported. In *panel b*, quantitative data are reported (means ± SD, $n = 2$ parallel samples; ***$p < 0.001$, as compared to untreated cells; ###$p < 0.001$, as compared to cells treated with CDDP only; two-sided, unpaired Student's t test)

4   Notes

1. To ensure the performance of the instrument over time, it is strongly recommended to periodically monitor the flow rate, laser alignment, and fluorescence stability. We also suggest to align/calibrate the instrument with standard beads, as per manufacturer's recommendations, before the beginning of each experimental session.

2. Based on the recommendations of the American Type Culture Collection (ATCC, Manassas, VA, US).

3. EDTA is not currently listed as a carcinogen by the International Agency for Research on Cancer (IARC), the National Toxicology Program (NTP) or the Occupational Safety and Health Administration (OSHA), yet it may be harmful if swallowed, and may behave as a mild irritant for the eyes, skin, and respiratory system.

4. Trypsin/EDTA solution is stable for at least 18 months, under appropriate storage conditions (−20 °C, under protection from light). The manufacturer recommends to avoid repeated freeze-thawing and to store trypsin/EDTA solution in aliquots of 2–10 mL. Once thawed, the product is stable at 4 °C for approximately 2 weeks.

5. CDDP is a platinum-based alkylating compound exerting clinical activity against a wide spectrum of solid neoplasms. The principal (but not the sole) mechanism of action of CDDP involves the generation of intra-/inter-strand DNA crosslinks and DNA-protein crosslinks [37, 38].

6. CDDP is currently listed as a carcinogen by IARC, NTP, or OSHA. CDDP is also mutagenic, fatal if swallowed, harmful if inhaled, or absorbed through skin and may cause serious eye damage. For these reasons, it should always be handled by wearing appropriate protective equipment (gloves, clothing, and eyewear).

7. Undissolved CDDP appears as a yellow to orange powder and has a minimum shelf life of >2 years if stored at 4 °C and protected from light. Under appropriate storage conditions (RT, vials sealed and protected from light), CDDP solutions are stable for several months.

8. DMF is not listed as a carcinogen by IARC, NTP, or OSHA, yet may be harmful in contact with eyes and skin, and is inflammable. Therefore, it should always be manipulated under a chemical fume hood and by using suitable protective equipment (gloves, clothing, and eyewear). In addition, it should be kept away from sources of ignition and stored tightly closed in a dry, well-ventilated place.

9. Z-VAD-fmk is a cell-permeable, irreversible pan-caspase inhibitor widely used in preclinical as a means to block the execution of regulated forms of cell death [15].

10. Z-VAD-fmk is not currently listed as a carcinogen by IARC, NTP, or OSHA, yet may behave as an irritant for the eyes, skin, and respiratory system.

11. Undissolved Z-VAD-fmk is stable for at least 2 years if stored at –20 °C, sealed, and protected from light. Under appropriate storage conditions (–20 °C, under protection from light), Z-VAD-fmk stock solutions are stable for at least 1 year. It is recommendable to store the reconstituted inhibitor in small aliquots (10–50 µL) and avoid repeated freeze-thawing.

12. DMSO is not listed as a carcinogen by IARC, NTP, or OSHA, yet may behave as irritant, sensitizer, may be harmful in contact with eyes, and is inflammable. Therefore, it should always be handled under a chemical fume hood, and by wearing appropriate protective equipment (gloves, clothing, and eyewear). In addition, it should always be kept away from sources of ignition and stored tightly closed in a dry, well-ventilated place.

13. $DiOC_6(3)$ is a cell-permeant, lipophilic dye that selectively accumulates within energized mitochondria, when used at low concentrations. $DiOC_6(3)$ exhibits excitation/emission peaks at 482/504 nm, respectively.

14. $DiOC_6(3)$ is not currently listed as a carcinogen by IARC, NTP, or OSHA, yet may behave as an irritant for the eyes, skin, and respiratory system. No mutagenic effects have been reported.

15. Under appropriate storage conditions (-20 °C, under protection from light) $DiOC_6(3)$ stock solution is stable for at least 12 months. Prolonged and unnecessary exposure to light of the solution should be minimized to prevent photobleaching.

16. PI is an exclusion dye, i.e., it penetrates into cells that have lost plasma membrane integrity. PI exhibits excitation/emission peaks at $482/504$ nm in aqueous solution and at $535/617$ nm, when bound to DNA.

17. PI is not currently listed as a carcinogen by IARC, NTP, or OSHA. This product, however, may have mutagenic effects, is harmful if swallowed, and behaves as irritant to eyes, respiratory system, and skin. Handling PI with the maximal care by wearing protective equipment (gloves, clothing, and eyewear) is recommended.

18. Undissolved PI has a dark red solid appearance and is stable for at least 12 months under standard storage conditions (RT, protected from light). In solution, PI is stable for at least 6 months (at 4 °C, under protection from light).

19. FITC-AnnV exhibits excitation/emission peaks at $494/518$ nm.

20. This agent contains sodium azide (NaN_3), which is toxic by ingestion and a severe irritant to the eyes and skin. Moreover, under acidic conditions NaN_3 yields the highly toxic gas hydrazoic acid. For these reason, FITC-AnnV should be always handled under a fume hood wearing appropriate equipment (gloves, clothing, and eyewear).

21. FITC-AnnV should be carefully protected from light to prevent photobleaching and should never be subjected to steep temperature changes. This product is stable for at least 6 months, under appropriate storage conditions (4 °C, under protection from light).

22. Under ordinary conditions of use and storage conditions this buffer is stable for at least 6 months. The manufacturer recommends to not freeze this product.

23. When large quantities of cells are needed or the physical space within incubators is limited alternative cell culture supports (i.e., 25 cm² or 175 cm² flasks) can be used.

24. It is recommendable to keep cells in culture by avoiding excessive underconfluence, which may result in a significant genetic drift of the population, as well as overconfluence, which may affect cell growth and viability by inducing metabolic, nutritional, and/or redox perturbations (*see also* **Notes 31** and **32**).

25. Cells are washed with PBS to remove traces of serum that may inactivate trypsin and thereby inhibit detachment. Prolonged or harsh washing should be avoided as it may provoke a sizeable loss of cells, especially for cell type that are particularly prone to detachment (such as HCT 116 cells).

26. Trypsin/EDTA can be replaced by TrypLE™ Express, a recombinant trypsin-like proteolytic enzyme that, as opposed to trypsin, can be stored for long periods at 4 °C, displays elevated stability at RT, and does not require inactivation during subculturing.

27. The incubation time for achieving complete detachment with trypsin/EDTA may vary with cell type and culture conditions. The majority of adherent cancer cells are efficiently detached by incubating cells for 1–3 min at 37 °C.

28. Over-trypsinization should be avoided, as it can result in severe cellular damage (e.g., degradation or internalization of growth factor receptors).

29. Detachment should be verified on light microscopy upon slight agitation of flasks/plates. Hitting or excessive shaking may result in cell clumps, which can be easily dissolved by repeatedly pipetting the cell suspension.

30. The serum contained in complete culture medium neutralizes trypsin.

31. According to ATCC, the standard subcultivation ratio for HCT 116 cells is 1:3–1:8. In our personal experience, HCT 116 cells can be safely diluted 1:15 without noticeable behavioral shifts in the population (*see also* **Notes 24** and **32**).

32. To preserve the stability of cultured cells, they should be always maintained in the exponential growth phase, and subcultured for a limited, predetermined number of passages (*see also* **Notes 24** and **31**). This requires a large stock of cryovials generated with early passage cells.

33. It is recommendable to allow recently thawed cells to readapt for at least two passages before employing them in experimental determinations.

34. The amount of cells that should be seeded may vary with cell type and specific experimental setting. 8×10^5 cells per well is usually appropriate for HCT 116 cells that will be analyzed 48 h upon seeding.

35. Some cells may require longer adaptation times upon seeding.

36. Confluence and general status of the cells should be carefully checked by light microscopy before any experimental intervention, as these parameters may affect both the uptake of some chemicals and the ability of cells to respond to them. As a general rule, confluence at treatment should not exceed 50 %.

37. To avoid the unwarranted loss of attached cells, it is recommended to remove the culture supernatant and replace it with the maximal care.

38. Untreated cells and cells exposed to equivalent amounts of solvent(s) provide an appropriate set of negative control conditions.

39. As an alternative to CDDP, the well-established inducer of caspase-dependent RCD staurosporine (final concentration = 0.1–1 μM) can be employed.

40. Samples should be checked on light microscopy for confluence and general status before collection.

41. Some instances of cell death proceed along with the detachment of cells from the culture substrate. Supernatants should therefore not be discarded unless the experimental settings are designed as to include only viable cells.

42. As general rule, all fluorochrome-containing solutions should be thoroughly mixed before use and maintained protected from light, in order to eliminate precipitates and avoid photobleaching, respectively.

43. When used at low concentrations, $DiOC_6(3)$ quickly accumulates within healthy, energized mitochondria in a virtually nonsaturatable fashion. To uniformly stain all samples of the same experiment, we suggest these general guidelines: (1) supernatants should be removed completely (by aspiration followed by inversion of FACS tubes on paper); (2) a single staining solution should be employed for all samples; (3) the staining solution must be repeatedly mixed throughout the experiment; (4) an equal volume of staining solution should be employed for all samples; and (5) cell clumps should be avoided (*see also* **Note 29**).

44. When used at high concentrations, $DiOC_6(3)$ may be display considerable quenching and stain the endoplasmic reticulum as well as other intracellular compartments.

45. Tetramethylrhodamine methyl ester (TMRM, excitation/emission peaks: 543/573 nm) at the final concentration of 150 nM can be employed as an alternative to $DiOC_6(3)$. TMRM, however, is spectrally incompatible with PI (*see also* **Note 48**).

46. Prolonged incubation with $DiOC_6(3)$ (>40 min) may be too toxic for some cells, while a short staining time (<20 min) may result in nonhomogenous or incomplete staining. Preliminary experiments aimed at optimizing staining conditions are suggested to avoid the under- or overestimation of dying cells (*see also* **Note 49**).

47. As an alternative, PI can be added to the $DiOC_6(3)$ staining solution at the final concentration of 0.5–1 μg/mL (*see* Subheading 3.2, **steps 5** and **6**). In this case, preliminary experiments to evaluate PI-dependent toxicity are required (*see also* **Note 49**).

48. As an alternative to PI, 10 μM 4′,6-diamidino-2-phenylindole (DAPI) can be employed (excitation/emission peaks at 358/461 nm, respectively). DAPI is compatible with both $DiOC_6(3)$ and TMRM (*see also* **Note 45**).

49. It should be noted that $DiOC_6(3)$ fluorescence may be influenced by $\Delta\psi_m$-independent, mitochondrion-dependent parameters (e.g., mitochondrial mass) as well as by mitochondrion-independent variables (e.g., cell size and plasma membrane potential). To estimate $\Delta\psi_m$-independent $DiOC_6(3)$ fluorescence, preliminary experiments should be performed whereby samples are divided in two aliquots prior to $DiOC_6(3)$ staining, one of which is exposed to 50–100 μM carbonyl cyanide *m*-chlorophenylhydrazone (CCCP, an ionophore that irreversibly dissipates $\Delta\psi_m$) for 5–10 min under standard conditions (37 °C, 5 % CO_2).

50. FSC reflects cell size, while SSC depends on the so-called refractive index, which is associated to various parameters including cell shape and granularity.

51. In our experimental setting, the following channels are employed for fluorescence detection: FL1 for $DiOC_6(3)$ and FL3 for PI.

52. If the experiment involves >30 specimens, we recommend to perform the staining and acquisition on <24 samples at a time, to control the exposure of cells to $DiOC_6(3)$ and PI (*see also* **Notes 46** and **47**).

53. Flow cytometers are normally equipped with software for instrumental control and first-line statistical analyses. Alternatively, various software for the analysis of cytofluorometric data is available online (for a catalog, refer to http://www.cyto.purdue.edu/flowcyt/software.htm).

54. To ensure statistical power, it is strongly recommended to acquire and analyze a large number of events exhibiting normal FSC and SSC (at least 10,000 events per sample).

55. Harsh or protracted washes may promote the exposure of PS on the cell surface, eventually favoring the overestimation of dying cells.

56. The FITC-AnnV solution should be prepared no early than 30 min before use and carefully shielded from light to limit photobleaching. FITC is significantly more sensitive to this problem than other dyes, including $DiOC_6(3)$ (*see also* **Notes 21** and **42**).

57. The total amount of FITC-AnnV necessary for optimal staining may vary with cell type and should therefore be estimated by preliminary experiments. It is not recommended to exceed the manufacturer's recommendations (i.e., 10 µL of FITC-AnnV in 100 µL of 1× cold (4 °C) binding buffer per 10^6 cells), as this may increase the likelihood of unspecific staining.

58. As an alternative, labeled samples may be kept for 30–45 min on ice under protection from light.

59. In our experimental setting, the following channels are employed for fluorescence detection: FL1 for FITC and FL3 for PI.

Acknowledgments

Authors are supported by the Ligue contre le Cancer (équipe labelisée); Agence National de la Recherche (ANR); Association pour la recherche sur le cancer (ARC); Cancéropôle Ile-de-France; AXA Chair for Longevity Research; Institut National du Cancer (INCa); Fondation Bettencourt-Schueller; Fondation de France; Fondation pour la Recherche Médicale (FRM); the European Commission (ArtForce); the European Research Council (ERC); the LabEx Immuno-Oncology; the SIRIC Stratified Oncology Cell DNA Repair and Tumor Immune Elimination (SOCRATE); the SIRIC Cancer Research and Personalized Medicine (CARPEM); the Paris Alliance of Cancer Research Institutes (PACRI); the Associazione Italiana per la Ricerca sul Cancro (AIRC, MFAG 2013 and Triennial Fellowship "Antonietta Latronico", 2014); the Ministero Italiano della Salute (GR-2011-02351355); and the Programma per i Giovani Ricercatori "Rita Levi Montalcini" 2011.

References

1. Ting JP, Willingham SB, Bergstralh DT (2008) NLRs at the intersection of cell death and immunity. Nat Rev Immunol 8:372–379

2. Maekawa T, Kufer TA, Schulze-Lefert P (2011) NLR functions in plant and animal immune systems: so far and yet so close. Nat Immunol 12:817–826

3. Kufer TA, Sansonetti PJ (2011) NLR functions beyond pathogen recognition. Nat Immunol 12:121–128

4. Krishnaswamy JK, Chu T, Eisenbarth SC (2013) Beyond pattern recognition: NOD-like receptors in dendritic cells. Trends Immunol 34:224–233

5. Zitvogel L, Kepp O, Galluzzi L, Kroemer G (2012) Inflammasomes in carcinogenesis and anticancer immune responses. Nat Immunol 13:343–351

6. Ghiringhelli F, Apetoh L, Tesniere A, Aymeric L, Ma Y, Ortiz C et al (2009) Activation of the NLRP3 inflammasome in dendritic cells induces IL-1beta-dependent adaptive immunity against tumors. Nat Med 15:1170–1178

7. Kroemer G, Galluzzi L, Kepp O, Zitvogel L (2013) Immunogenic cell death in cancer therapy. Annu Rev Immunol 31:51–72

8. Proell M, Riedl SJ, Fritz JH, Rojas AM, Schwarzenbacher R (2008) The Nod-like receptor (NLR) family: a tale of similarities and differences. PLoS One 3, e2119

9. Krysko DV, Garg AD, Kaczmarek A, Krysko O, Agostinis P, Vandenabeele P (2012) Immunogenic cell death and DAMPs in cancer therapy. Nat Rev Cancer 12:860–875

10. Galluzzi L, Kepp O, Kroemer G (2012) Mitochondria: master regulators of danger signalling. Nat Rev Mol Cell Biol 13:780–788

11. Garg AD, Martin S, Golab J, Agostinis P (2014) Danger signalling during cancer cell death: origins, plasticity and regulation. Cell Death Differ 21:26–38

12. Galluzzi L, Bravo-San Pedro JM, Kroemer G (2014) Organelle-specific initiation of cell death. Nat Cell Biol 16:728–736

13. Galluzzi L, Maiuri MC, Vitale I, Zischka H, Castedo M, Zitvogel L et al (2007) Cell death modalities: classification and pathophysiological implications. Cell Death Differ 14:1237–1243

14. Galluzzi L, Pietrocola F, Levine B, Kroemer G (2014) Metabolic control of autophagy. Cell 159:1263–1276

15. Galluzzi L, Bravo-San Pedro JM, Vitale I, Aaronson SA, Abrams JM, Adam D et al (2015) Essential versus accessory aspects of cell death: recommendations of the NCCD 2015. Cell Death Differ 22:58–73

16. Galluzzi L, Vitale I, Abrams JM, Alnemri ES, Baehrecke EH, Blagosklonny MV et al (2012) Molecular definitions of cell death subroutines: recommendations of the Nomenclature Committee on Cell Death 2012. Cell Death Differ 19:107–120

17. Green DR, Galluzzi L, Kroemer G (2011) Mitochondria and the autophagy-inflammation-cell death axis in organismal aging. Science 333:1109–1112

18. Galluzzi L, Aaronson SA, Abrams J, Alnemri ES, Andrews DW, Baehrecke EH et al (2009) Guidelines for the use and interpretation of assays for monitoring cell death in higher eukaryotes. Cell Death Differ 16:1093–1107

19. Kepp O, Galluzzi L, Lipinski M, Yuan J, Kroemer G (2011) Cell death assays for drug discovery. Nat Rev Drug Discov 10:221–237

20. Green DR, Galluzzi L, Kroemer G (2014) Cell biology. Metabolic control of cell death. Science 345:1250256

21. Tait SW, Green DR (2010) Mitochondria and cell death: outer membrane permeabilization and beyond. Nat Rev Mol Cell Biol 11: 621–632

22. Kroemer G, Galluzzi L, Brenner C (2007) Mitochondrial membrane permeabilization in cell death. Physiol Rev 87:99–163

23. Taylor RC, Cullen SP, Martin SJ (2008) Apoptosis: controlled demolition at the cellular level. Nat Rev Mol Cell Biol 9:231–241

24. Galluzzi L, Kepp O, Krautwald S, Kroemer G, Linkermann A (2014) Molecular mechanisms of regulated necrosis. Semin Cell Dev Biol 35:24–32

25. Vanden Berghe T, Linkermann A, Jouan-Lanhouet S, Walczak H, Vandenabeele P (2014) Regulated necrosis: the expanding network of non-apoptotic cell death pathways. Nat Rev Mol Cell Biol 15:135–147

26. Linkermann A, Green DR (2014) Necroptosis. N Engl J Med 370:455–465

27. Zamzami N, Marchetti P, Castedo M, Zanin C, Vayssiere JL, Petit PX et al (1995) Reduction in mitochondrial potential constitutes an early irreversible step of programmed lymphocyte death in vivo. J Exp Med 181:1661–1672

28. Zamzami N, Marchetti P, Castedo M, Decaudin D, Macho A, Hirsch T et al (1995) Sequential reduction of mitochondrial transmembrane potential and generation of reactive oxygen species in early programmed cell death. J Exp Med 182:367–377

29. Galluzzi L, Zamzami N, de La Motte Rouge T, Lemaire C, Brenner C, Kroemer G (2007) Methods for the assessment of mitochondrial membrane permeabilization in apoptosis. Apoptosis 12:803–813

30. Segawa K, Kurata S, Yanagihashi Y, Brummelkamp TR, Matsuda F, Nagata S (2014) Caspase-mediated cleavage of phospholipid flippase for apoptotic phosphatidylserine exposure. Science 344:1164–1168

31. Cai Z, Jitkaew S, Zhao J, Chiang HC, Choksi S, Liu J et al (2014) Plasma membrane translocation of trimerized MLKL protein is required for TNF-induced necroptosis. Nat Cell Biol 16:55–65

32. Wang H, Sun L, Su L, Rizo J, Liu L, Wang LF et al (2014) Mixed lineage kinase domain-like protein MLKL causes necrotic membrane

disruption upon phosphorylation by RIP3. Mol Cell 54:133–146

33. Degterev A, Huang Z, Boyce M, Li Y, Jagtap P, Mizushima N et al (2005) Chemical inhibitor of nonapoptotic cell death with therapeutic potential for ischemic brain injury. Nat Chem Biol 1:112–119

34. Wallach D (2005) Probing cell death by chemical screening. Nat Chem Biol 1:68–69

35. Degterev A, Hitomi J, Germscheid M, Ch'en IL, Korkina O, Teng X et al (2008) Identification of RIP1 kinase as a specific cellular target of necrostatins. Nat Chem Biol 4:313–321

36. Slee EA, Zhu H, Chow SC, MacFarlane M, Nicholson DW, Cohen GM (1996) Benzyloxycarbonyl-Val-Ala-Asp (OMe) fluoromethylketone (Z-VAD.FMK) inhibits apoptosis by blocking the processing of CPP32. Biochem J 315(Pt 1):21–24

37. Galluzzi L, Vitale I, Michels J, Brenner C, Szabadkai G, Harel-Bellan A et al (2014) Systems biology of cisplatin resistance: past, present and future. Cell Death Dis 5, e1257

38. Galluzzi L, Senovilla L, Vitale I, Michels J, Martins I, Kepp O et al (2012) Molecular mechanisms of cisplatin resistance. Oncogene 31:1869–1883

Chapter 18

NLR in Human Diseases: Role and Laboratory Findings

Sonia Carta, Marco Gattorno, and Anna Rubartelli

Abstract

Autoinflammatory diseases are a group of inherited and multifactorial disorders characterized by an overactivation of innate immune response. In most cases, the clinical manifestations are due to increased activity of the NLRP3 inflammasome resulting in increased IL-1β secretion. Investigating inflammatory cells from subjects affected by autoinflammatory diseases presents a number of technical difficulties related to the rarity of the diseases, to the young age of most patients, and to the difficult modulation of gene expression in primary cells. However, since cell stress is involved in the pathophysiology of these diseases, the study of freshly drawn blood monocytes from patients affected by IL-1-mediated diseases strongly increases the chances that the observed phenomena is indeed pertinent to the pathogenesis of the disease and not influenced by the long-term cell culture conditions (e.g., the high O_2 tension) or gene transfection in continuous cell lines that may lead to artifacts.

Key words Autoinflammatory diseases, Primary monocytes, ATP, IL-1β secretion, Redox

1 Introduction

Autoinflammatory diseases (AID) are a group of multisystem disorders characterized by recurrent episodes of fever and systemic inflammation affecting the eyes, joints, skin, and serosal surfaces. These syndromes differ from autoimmune diseases by several features, including the periodicity whereas autoimmune diseases are progressive, and the lack of signs of involvement of adaptive immunity such as association with HLA aplotypes, high-titer autoantibodies, or antigen-specific T cells. Thus, autoinflammatory syndromes are recognized as disorders of innate immunity [1]. This definition is supported by the dramatic therapeutic response to IL-1 blocking. Indeed, the rapid and sustained response to a reduction in IL-1 activity on an "ex adjuvantibus" basis is the best hallmark of most of these diseases [2, 3].

Due to the rarity of these conditions, most of the studies aimed to unravel the pathogenic consequences related to the mutation of genes involved in inherited autoinflammatory diseases were based on the analysis of in vitro transfected cells or animal models.

Francesco Di Virgilio and Pablo Pelegrín (eds.), *NLR Proteins: Methods and Protocols*, Methods in Molecular Biology, vol. 1417, DOI 10.1007/978-1-4939-3566-6_18, © Springer Science+Business Media New York 2016

These approaches have the clear advantage to facilitate the availability of material for these studies and also to reduce the variability associated to clinical and genetic variables (type of mutation, active versus inactive disease, ongoing treatment, individual responses to stress, etc.). On the other hand the use of patients' primary cells strongly increase the possibility that the observed phenomena could be indeed pertinent to the pathogenesis of the disease and not influenced by possible artifacts linked to the study of transfected cells or animal models.

In this chapter, we describe the basic laboratory procedures for the investigation of NLR and inflammasome activation in primary cells from individuals affected by inherited autoinflammatory diseases.

2   Materials

2.1   Isolation of Peripheral Blood Mononuclear Cells (PBMCs)

1. Ficoll-Paque.
2. Heparinized tubes to collect blood.

2.2   PBMC Characterization

1. Phycoerythrin-conjugated anti-CD14 monoclonal antibody.
2. Flow cytometer (e.g., Becton-Dickinson Canto) or equivalent with appropriate lasers and detectors.

2.3   PBMC Culture

1. Humidified 5 % CO_2 and 37 °C incubator.
2. Cell culture plates of different formats (96-, 24-, or 6-well plates).
3. RPMI-FBS: Roswell Park Memorial Institute (RPMI) 1640 medium containing 5 % fetal bovine serum (*see* **Note 1**).
4. RPMI-HU: RPMI 1640 medium supplemented with 1 % Nutridoma-HU (Roche Applied Science) (*see* **Note 1**).

2.4   PBMC Agonists and Inhibitors

1. Lipopolysaccharide (LPS) from *Escherichia coli* 0111:B4.
2. Selective TLR7 ligand R848.
3. Zymosan.
4. Diphenylene iodonium, DPI (*see* **Note 2**).
5. 1,3-bis(2-chloroethyl)-1-nitrosourea, BCNU (*see* **Note 3**).
6. Adenosine triphosphate, ATP.
7. Oxidized ATP, oATP (*see* **Note 4**).
8. 6-N,N-diethyl-d-β-γ-dibromomethylene adenosine triphosphate, ARL67156 (*see* **Note 5**).

2.5 Transfection, RNA Isolation, and Amplification	1. TriPure Isolation Reagent (Roche) or other similar reagent that allows RNA isolation from cell samples. 2. NLRP3 predesigned short interfering RNA. 3. Amaxa Nucleofector™ Technology (Lonza). 4. Lymphocyte Growth Media-3 LGM-3, IMDM media (Lonza) (*see* **Note 6**). 5. Reverse Transcription Kit. 6. Platinum SYBR green qPCR SuperMix-UDG (Invitrogen) or equivalent.
2.6 Measurement of IL-1β Release	1. Human IL-1β ELISA kit. 2. Trichloroacetic acid (TCA). 3. Acetone.
2.7 Measurement of Intracellular Reactive Oxygen Species (ROS)	1. 2,7-Dichlorodihydrofluorescein diacetate, H2DCF-DA. 2. 0.2 % Triton X-100. 3. Detergent compatible (DC) protein assay.
2.8 Measurement of Cysteine Release	1. 5,5-Dithiobis-2-nitrobenzoic acid, DTNB. 2. Cysteine and cystine. 3. Oxidized glutathione, GSH. 4. Bondapak NH_2 column.
2.9 Measurement of ATP Release	1. RPMI-FCS (*see* **Note 1**). 2. ATP Bioluminescence Assay Kit HS II (Roche) or equivalent.

3 Methods

3.1 Isolation of Primary Monocytes from Patients with AID

1. PBMC fraction is obtained by differential centrifugation of freshly drawn heparinized blood over Ficoll-Paque gradients (*see* **Note 7**). PBMC are stained with phycoerythrin-conjugated anti-CD14 monoclonal antibody and analyzed by flow cytometer to determine the percentage of monocytes. In healthy subjects, the percentage of monocytes ranges between 10 and 20 % [4].

2. PBMC are adjusted to 10^7/ml in RPMI-FBS medium (*see* **Note 1**) and plated in 96-well (0.1 ml/well), 24-well (0.4 ml/well), or 6-well (2 ml/well) plates.

3. Monocytes are enriched by adherence by 45–60 min incubation at 37 °C and 5 % CO_2.

4. Non adherent cells are discharged by removing the cell culture media.

3.2 Culture Conditions and Stimulation of Enriched Monocytes Preparations from Patients with AID

1. Adherent cells are stimulated with various doses of LPS (from 100 ng/ml to 10 pg/ml) [4, 5], R848 (5 µg/ml) or zymosan (20 µg/ml) [6] for different times (from 3 to 18 h) at 37 °C and 5 % CO_2 in either RPMI-FBS or RPMI-HU [7–9] (*see* **Notes 1** and **8**).

2. Secretion of IL-1β is used as readout for NLRP3 inflammasome activation. To inhibit inflammasome activation and IL-1β secretion, monocytes are treated with DPI (20 µM), BCNU (50 µM) [5], oATP (300 µM), or ARL67156 (200 µM) for different times (3–18 h) [4].

3. At the end of the incubation, supernatants are collected for IL-1β or cysteine determination (*see* Subheading 3.6). To study the effect of exogenous ATP on IL-1β secretion, monocytes (*see* Subheading 3.1, **step 2**) in RPMI-FBS or RPMI-HU are primed with LPS for 3 h, supernatants are withdrawn and replaced with RPMI-FBS or RPMI-HU (*see* **Note 8**) supplemented with or without 1 mM ATP, for 20 min at 37 °C and 5 % CO_2. Supernatants are collected for IL-1β determination.

4. IL-1β secreted in the supernatants is determined by ELISA or immunoblot as described in Subheading 3.3.

5. Cells are then lysed in Triton X-100 lysis buffer for western blot analysis or in TriPure Isolation Reagent for RNA extraction [4, 10, 11].

3.3 Analysis of Pro- and Mature IL-1β Production and Secretion

1. Supernatants from adherent cells cultured in RPMI-HU are transferred to 1.5 ml tubes.

2. The protease inhibitor PMSF (1 mM) is added.

3. Supernatants are centrifuged at $9300 \times g$ for 5 min at 4 °C in a minifuge to eliminate cell debris.

4. TCA is added to supernatants to precipitate proteins (10 % vol/vol).

5. Samples are incubated 1 h at room temperature.

6. Precipitated proteins are pelleted at $13,400 \times g$ for 10 min at 4 °C in a minifuge.

7. Pellets are washed 3× with acetone ($13,400 \times g$ for 10 min).

8. The dry pellets are resuspended in H_2O with 1 % Tris base and 1 mM PMSF.

9. Equal amounts of 2× reducing Laemmli sample buffer are added.

10. Samples are heated for 5 min at 95 °C [10].

3.4 NLRP3 mRNA Silencing in Primary Cells

1. NLRP3 mRNA silencing is performed on PBMC, isolated from heparinized blood of healthy individuals or AID patients as described in Subheading 3.1 using the Human Monocytes Nuclofector kit. PBMC (3×10^6–1×10^7 cells) are resuspended

in 100 µl of Human Monocytes Nuclofection Solution and mixed with 5 µg of mock or NLRP3 siRNA. The sample is transferred into an Amaxa cuvette and nucleofected using the program Y-001 for Nucleofector II Device [12].

2. Post-electroporation, cells are immediately transferred to culture plates containing prewarmed culture medium (*see* **Note 6**) at a density of 5×10^5 cells/well (for 96-well plate) or 2×10^6 cells/well (for 24-well plate) [12].

3. Medium is changed after 1 h (*see* Subheading 3.1) and after 4 or 24 h (*see* **Note 9**) adherent cells are stimulated with 1 µg/ml of LPS for 18 h. The level of IL-1β is measured in the supernatants at the end of incubation as an index of NLRP3 inflammasome activation (*see* **Note 10**) [13].

4. To validate silencing, cells are then lysed in Triton X-100 lysis buffer for western blot analysis or in TriPure Isolation Reagent for RNA extraction.

3.5 Evaluation of Intracellular ROS in Primary Cells from AID Patients

1. PBMCs are isolated from heparinized blood of healthy individuals or AID patients as described in Subheading 3.1.

2. Monocytes are stimulated with different TLR agonists (LPS 100 ng/ml, R848 5 µg/ml, or zymosan 20 µg/ml) for 1 h. Thirty min before the end of the incubation the fluorescent dye H_2DCF-DA (10 µM) is added to the cells.

3. Cells are then lysed in 0.2 % Triton-X100 and H_2DCF-DA fluorescence is measured in the cell lysates with a microplate fluorometer at the 480/530 nm excitation/emission wavelength pair.

4. Fluorescence signal intensity is normalized vs the protein content of each sample evaluated by DC protein assay [5].

3.6 Determination of Cysteine in Cellular Culture Media

1. Supernatants (0.1 ml) from monocytes cultured at $4 \times 10^5 / 0.4$ ml in 24-well plates (*see* Subheadings 3.1 and 3.2) are transferred to 96-well plate and reacted with 10 mM DTNB. The absorption is measured immediately using a microplate reader (e.g., ELx808 absorbance microplate reader BioTek) set to 412 nm. Cysteine is used as standard. A seven point standard curve using twofold serial dilution of cysteine (from 50 to 2.5 µM) in RPMI-FCS is utilized.

2. To discriminate between extracellular cysteine and GSH, high-performance liquid chromatography is used. *S*-carboxymethyl derivatives of soluble thiols were generated by reaction of free thiols with iodoacetic acid followed by conversion of free amino groups to 2,4-dinitrophenyl derivatives with 1-fluoro-2,4-dinitrobenzene. These derivatives are separated by HPLC with a Bondapak NH2 column. GSH, oxidized glutathione, cystine, and cysteine are used as external standards [5].

3.7 Determination of ATP Secretion from Freshly Isolated Monocytes

1. PBMCs are isolated from heparinized blood of healthy individuals or AID patients as described in Subheading 3.1 and then activated with LPS (*see* **Note 11**), for 3 h at 37 °C in 5 % CO_2 in RPMI-FCS, in the presence or absence of ARL67156 or DPI. After 3 h, cell supernatants were collected for ATP measurement.

2. For measurement of extracellular ATP concentration, 50 μl of cell supernatant are transferred to 96-white microtiter plate and 50 μl of luciferase reagent (ATP Bioluminescence Assay Kit HS II) are added. Luminescence is measured with a microplate luminometer (e.g., Luminoskan Ascent Thermo Electron Corporation) [4, 14]. Serial dilution of ATP in the range of 10^{-6} to 10^{-12} M is used as standard curve.

3.8 Evaluation of the Expression and Modulation of Redox Gene and Cytokine Gene in Adherent Cells by AID Patients by Real-Time PCR

1. Total mRNA is isolated from monocytes, unstimulated or stimulated with TLR agonists, with the TriPure Isolation Reagent and reverse-transcribed with the Reverse Transcription Kit.

2. Real-time PCR is performed using Platinum SYBR green qPCR SuperMix-UDG. The specific primers for IL-1β, NLRP3, xCT, and GAPDH are described in refs [12–15]. Target gene levels are normalized to those of GAPDH mRNA, and relative expression is determined using the ΔCt method [4].

4 Notes

1. PBMC are cultured in RPMI-FBS (RPMI 1640 supplemented with 2 mM l-glutamine, 5 % FBS, 100 U/ml penicillin, and 100 μg/ml streptomycin) or RPMI-HU (RPMI 1640 supplemented with 2 mM l-glutamine, 100 U/ml penicillin, and 100 μg/ml streptomycin and 1 % Nutridoma-HU).

2. DPI is an inhibitor of flavoproteins that prevents ROS production.

3. BCNU inhibits thioredoxin activity.

4. oATP is an irreversible inhibitor of the purinergic receptor P2X7.

5. ARL67156 is a specific ecto-ATPase inhibitor.

6. Transfected cells are cultured in Lymphocyte Growth Media-3 LGM-3 (serum-free culture) or IMDM media supplemented with 10 % FBS, 100 U/ml penicillin, 100 μg/ml streptomycin and with the addition of 20 ng/ml human recombinant γ-interferon.

7. Ficoll-Paque is a solution of high molecular weight sucrose polymers and sodium diatrizoate used for isolating mononuclear cells from blood using a centrifugation procedure. Ficoll-

Paque is placed at the bottom of a conical tube, and blood is then slowly layered above. After being centrifuged ($1800 \times g$ for 20 min), the following layers will be visible from top to bottom: plasma and other constituents, PBMC, Ficoll-Paque, and erythrocytes and granulocytes which are present in pellet form. PBMC layer is collected, placed in a new tube and rinsed $3\times$ with PBS (at $515 \times g$, $394 \times g$, and $290 \times g$ for 5 min).

8. Monocytes are cultured in RPMI-HU for Western Blot Analysis of supernatants.

9. The incubation period before stimulation with TLR agonist depends on the gene of interest. In some experiments, gene expression is often detectable after only 4–6 h. If this is the case, the cells are incubated in culture medium without human recombinant γ-interferon.

10. In these experimental conditions, the levels of secreted IL-1β are lower than those detected in supernatants of monocytes stimulated immediately after isolation [13].

11. Different concentrations of LPS can be used, from 0.001 to 100 ng/ml.

References

1. Masters SL, Simon A, Aksentijevich I, Kastner DL (2009) Horror autoinflammaticus: the molecular pathophysiology of autoinflammatory disease. Annu Rev Immunol 27:621–668
2. Dinarello CA (2011) Interleukin-1 in the pathogenesis and treatment of inflammatory diseases. Blood 117:3720–3732
3. Gattorno M, Martini A (2013) Beyond the NLRP3 inflammasome: autoinflammatory diseases reach adolescence. Arthritis Rheum 65:1137–1147
4. Carta S, Penco F, Lavieri R, Martini A, Dinarello CA, Gattorno M et al (2015) Cell stress increases ATP release in NLRP3 inflammasome-mediated autoinflammatory diseases, resulting in cytokine imbalance. Proc Natl Acad Sci U S A 112:2835–2840
5. Tassi S, Carta S, Delfino L, Caorsi R, Martini A, Gattorno M et al (2010) Altered redox state of monocytes from cryopyrin-associated periodic syndromes causes accelerated IL-1beta secretion. Proc Natl Acad Sci U S A 107:9789–9794
6. Carta S, Tassi S, Delfino L, Omenetti A, Raffa S, Torrisi MR et al (2012) Deficient production of IL-1 receptor antagonist and IL-6 coupled to oxidative stress in cryopyrin-associated periodic syndrome monocytes. Ann Rheum Dis 71:1577–1581
7. Andrei C, Dazzi C, Lotti L, Torrisi MR, Chimini G, Rubartelli A (1999) The secretory route of the leaderless protein interleukin 1beta involves exocytosis of endolysosome-related vesicles. Mol Biol Cell 10:1463–1475
8. Andrei C, Margiocco P, Poggi A, Lotti LV, Torrisi MR, Rubartelli A (2004) Phospholipases C and A2 control lysosome-mediated IL-1 beta secretion: implications for inflammatory processes. Proc Natl Acad Sci U S A 101:9745–9750
9. Carta S, Tassi S, Semino C, Fossati G, Mascagni P, Dinarello CA et al (2006) Histone deacetylase inhibitors prevent exocytosis of interleukin-1beta-containing secretory lysosomes: role of microtubules. Blood 108:1618–1626
10. Gattorno M, Tassi S, Carta S, Delfino L, Ferlito F, Pelagatti MA et al (2007) Pattern of interleukin-1beta secretion in response to lipopolysaccharide and ATP before and after interleukin-1 blockade in patients with CIAS1 mutations. Arthritis Rheum 56:3138–3148
11. Gattorno M, Piccini A, Lasiglie D, Tassi S, Brisca G, Carta S et al (2008) The pattern of response to anti-interleukin-1 treatment distinguishes two subsets of patients with systemic-onset juvenile idiopathic arthritis. Arthritis Rheum 58:1505–1615
12. Carta S, Tassi S, Pettinati I, Delfino L, Dinarello CA, Rubartelli A (2011) The rate of interleukin-

1beta secretion in different myeloid cells varies with the extent of redox response to Toll-like receptor triggering. J Biol Chem 286:27069–27080

13. Omenetti A, Carta S, Delfino L, Martini A, Gattorno M, Rubartelli A (2014) Increased NLRP3-dependent interleukin 1beta secretion in patients with familial Mediterranean fever: correlation with MEFV genotype. Ann Rheum Dis 73:462–469

14. Lavieri R, Piccioli P, Carta S, Delfino L, Castellani P, Rubartelli A (2014) TLR costimulation causes oxidative stress with unbalance of proinflammatory and anti-inflammatory cytokine production. J Immunol 192:5373–5381

15. Tassi S, Carta S, Vene R, Delfino L, Ciriolo MR, Rubartelli A (2009) Pathogen-induced interleukin-1beta processing and secretion is regulated by a biphasic redox response. J Immunol 183:1456–1462

ERRATUM TO

Chapter 9
Measuring NLR Oligomerization II: Detection of ASC Speck Formation by Confocal Microscopy and Immunofl uorescence

Michael Beilharz, Dominic De Nardo, Eicke Latz, and Bernardo S. Franklin

Francesco Di Virgilio and Pablo Pelegrín (eds.), *NLR Proteins: Methods and Protocols*, Methods in Molecular Biology,
vol. 1417, DOI 10.1007/978-1-4939-3566-6_9, © Springer Science+Business Media New York 2016

DOI 10.1007/978-1-4939-3566-6_19

Due to a typesetting error, The author name Dominic De Nardo was pluralized and was wrong in the initial version published online and in print. The correct name, as appears now in both the print and online version of the book, is:

Michael Beilharz, Dominic De Nardo, Eicke Latz, and Bernardo S. Franklin

The online version of the original chapter can be found at
http://dx.doi.org/10.1007/978-1-4939-3566-6_9

INDEX

Francesco Di Virgilio and Pablo Pelegrín (eds.), *NLR Proteins: Methods and Protocols*, Methods in Molecular Biology,
vol. 1417, DOI 10.1007/978-1-4939-3566-6, © Springer Science+Business Media New York 2016